Population Studies

THE AUTHOR

Dr. Vijaya P. Khairkar is a renowned population geographer whose research focuses on the intricate relationships between demographic changes and spatial patterns. With a Ph.D. in Geography, she has dedicated her career to studying population dynamics, migration, and urbanization, contributing valuable insights to the field. Her work emphasizes the importance of understanding population trends in addressing social and environmental challenges. Dr. Khairkar has published numerous articles and books, and she actively engages in academic discourse through conferences and workshops. Passionate about teaching, she inspires her students to think critically about the implications of population studies in today's world. Her commitment to integrating theory with practical applications makes her a respected figure in both academic and policy-making circles.

POPULATION STUDIES

Author

Dr. Vijaya P. Khairkar

ACADEMIC
ENCLAVE

First Edition, 2025

ISBN: 978-81-971898-6-9 (Hardback)

Printed at Replika Press Pvt. Ltd.

Published by
ACADEMIC ENCLAVE
House No. 10/15, Ground Floor,
Block-A, Street No. 14/3, Kamal Vihar,
Kamal Pur, Burari, Delhi-110084
Email: info@academicenclave.com
Mobile: +91-7042423166
Website: www.academicenclave.com

Preface

"Population Studies" offers a comprehensive exploration of the dynamics shaping human populations across the globe. In an era marked by rapid demographic changes, understanding these shifts is crucial for policymakers, researchers, and engaged citizens alike. This book delves into the intricate interplay of factors such as migration, mortality, fertility, and nuptiality, while also addressing the socio-economic implications of population trends.

This book combines theoretical insights with empirical research, offering a comprehensive overview of contemporary issues in population studies. Each chapter is crafted to engage readers, encouraging critical thinking and fostering an appreciation for the interdisciplinary nature of this field. It is my hope that this work not only enhances academic discourse but also inspires action towards creating equitable and sustainable societies.

I extend my heartfelt gratitude to the contributors whose expertise enriches this work and to the readers who embark on this journey of exploration with us. It is my hope that this book will serve as a valuable resource for students, scholars, and practitioners alike, inspiring informed dialogue and innovative solutions to the challenges posed by population changes in our rapidly evolving world.

Dr. Vijaya P. Khairkar

Contents

Preface v

1. **Population Concept and Theories** 1
   - Origin and Development of Population Geography 3
   - Difference Between Population Geography and Demography 5
   - Approaches to the Study of Population Geography 6
   - Sources of Population Data 6
   - Major Sample Surveys in India 7
   - Population Theories 7
   - Religion and Population Issues 9
   - Optimum Population Theory 12
   - The Theory of Demographic Transition 14
   - Demographic Transition Theory and Criticism 16
2. **Population Composition** 17
   - Population Parameter for India 2011 19
   - Age Composition in India 20
   - Sex Ratio of India 21
   - Caste Composition in India 2011 22
   - Literacy in India 22
   - Millennium Development Goals (MDGs) 23
3. **Fertility** 25
   - Introduction 25
   - Basic Concepts in Fertility Studies 25
   - Sources of Data 26
   - Theory of Diffusion or Cultural Lag 27
   - Becker's Theory 30
   - Measures of Fertility 34
   - Fertility Situation in India 40

4. Mortality **41**
- Sources of Data 42
- The Expectation of Life at Birth 45
- Endogenous Factors 45
- Mortality in India 47

5. Migration **49**
- Measures of Migration 50
- Types of Migration 50
- Push-Pull Hypothesis 51

6. Marital Status (Nuptiality) **55**
- Introduction 55
- Sources and Quality of Data Sources of data: 57
- Measures of Nuptiality 58
- Concepts and Definition of Terms 58
- India's Case 63
- Interpretation of Age at Marriage for Women in India (1992-2022) 70
- Social and Economic Factors 71

7. Population Policies **73**
- India's Population Policy 73
- The Family Planning Programs 77
- The Extension Approach 80
- The Camp Approach 83
- The Reproductive and Child Health Approach 84
- Reproductive Health 85
- Population and Sustainable Development Goals 87

References **91**

Colour MAPS **93**

CHAPTER-1

Population Concept and Theories

The population is a dynamic phenomenon. Population refers to the total number of humans residing in the entire world. The estimate for India's population as of October 8, 2024, is 1,454,427,269. This makes India the most populated country in the world, accounting for 17.78 percent of the world's total population. Today, the population of the world is more than 8 billion. Now, India crosses 150 crore population. Population consists of fertility, mortality, migration, and nuptiality. Changes in one factor affecting population growth, structure, and composition. Hence, studying population phenomena is vital part in the study of human kind. A population is defined as a group of individuals of the same species existing and crossing within a given area. The most vital demographic parameter is the number of individuals within a population (Lebreton, et al. 1992).

Geography is a study of landscape and distribution of physical and cultural factors over the surface of the earth. The physical, biological, and social sciences have their own philosophy, methodology, and scope. Similarly geography numerous tangibles and in tangible natural and manmade phenomena. In human geography, the major force is on the study of human culture in relation to the habitation or environment the subject deal with the spatial-temporal distribution of societies, human geography covers a very wide field, or its scope is vast. It includes the study of human races, the growth, distribution, and density of populations in the different parts of the world, their demographic characteristics and migration patterns, and physical and cultural differences between human groups and economic occupations. It also covers the relationship between man and his natural environment and the way in which his actions are distributed. Human geography also considers the combination of culture, language, religion, customs and traditions type and patterns of rural settlements.

Population geography is a recently developed specialized sub-branch of human geography. The scope of this branch is so vast that today it is treated as an independent, specialized branch. Population itself is an important aspect that varies over the surface of the earth. In population geography, we study the distribution of population, its growth and characteristics, its religious and linguistic composition, sex ratio, age composition, migration, standard of living, its economic structure, etc. with respect to location. We also study the adaptation of several human groups to their respective environments in different parts of the world. Different aspects of population have been studied since long. According to

Donald Bogue, demography is the study of the size of population, its structure, characteristics, and changes in all these with the help of fertility, mortality, marriage, migration, and social mobility. According to Warren Thompson and David Lewis, demography is the study of characteristics of various human groups, their distribution, changes in population, and their effect in relation to the well-being of society. G.T. Trewartha confined the scope of population geography. Growth of population in the past, distribution of population, density, migration, growth, qualities of population, and their regional patterns of distribution. He divided his book, The Geography of Population, into two parts. The first deals with the geography of population in the past, and the second with all the characteristics of population. This means According to Trewartha; population geography studies the characteristics of various human groups found in different geographical regions. According to Clarke, population geography studies the characteristics of several human groups found in different geographical regions and their relationship with physical, cultural, and economic factors. According to Zelinsky, population geography means the study of the process through which the changing population with space and time exerts its influence on the geographical characteristics of the region and, at the same time, the impact of the geographical characteristics of the region on the population. In short, a precise and comprehensive definition of population geography could be, "Population geography studies the distribution of population over the surface of the earth along with its characteristics and its relation with the geographical personality of the region.". Population geography can also be defined as "the study of the distribution of population in various regions, its characteristics, the impact of the geographical personality of the region on various aspects of the population, and the adaptation of the people to their respective geographical surroundings and the impact of the population on the geography of the region. The following points will help elaborate on the scope and field of population geography. Distribution of population and their characteristics in different continents, sub-continents, countries, and states; nature of distribution; changes in the future and their impact. The density of population, its various types, physical and cultural factors, and the structure of population density in different parts of the world. Migration and international migration, factors affecting migration, types of migration, causes and consequences of migration, laws of migration, impact of geographical factors on migration, and vice versa. Growth of population: measure of growth of population, birth rate, death rate. Factors affecting these trends of birth rate and death rate in various parts and theories of population growth. Composition of population: composition of population on the basis of age, sex, education, religion, language, marital status, occupation, and economic status. Impact of geographical phenomena of the region on the composition of population and vice versa.

Literacy: proportion of literates in various parts of the world and relationship between quality of population and literacy.

Urbanization: rural population, urban population. Rural and urban populations have their own distinct characteristics. Their study is essential because it reveals the adaptation of these groups to their respective geographical surroundings. It is also essential to study the process of urbanization and trends of urbanization in various parts of the world. To study the relationship between population and natural resources and the level of technological development. Development of natural resources, relationship between development and growth of population, pressure of population on resources, the study of overpopulation and underpopulation and optimum population, the balance between population, resources, and the level of technological development in various parts of the world, and to formulate policies to complete the balance between geographical setting, resources, technological development, and population.

Origin and Development of Population Geography

Origin and Development of Population Geography: Though population geography has emerged as an independent discipline only recently, various aspects related to population have been studied for a long time. Discussion regarding how to enumerate economic and agricultural aspects of population is found in a book written by Kautillya (321-269 B.C.) on Economics. Reference to population and industries is also found in Aine-Akbari, written during the regime of Emperor Akbar. References to population are also found in the writings of Aristotle and Plato. The scientific study of population began from the seventeenth century onwards. John Grount (1620-1674) is considered the father of demography. He arrived at certain conclusions related to population by using birth and death records maintained by the churches in and around London. For example, the birth and death certificates in London are in the ratio of 11:12; deaths amongst males are higher than those amongst females; every year, 6,000 people come to London from rural areas, etc. Grount's work was continued by Gregory King and William Petty. The term average expectancy of life was first used by Edmond Halley. Mortality tables were prepared by Sasoo Milch, a German scientist, in his book written during the 17th century. Daniel Baraulli and Monian contributed towards the study of demography during the 18th and 19th centuries. Barnaulli arrived at certain conclusions by analyzing the data related to the inoculation of smallpox. Montian discussed the causes of population growth of Demography was further helped by the theories put forward by Malthus (1766-1834) and Karl Marx in France. Study of Demography further helped by the theories put forward by Malthus (1766-1834) and Karl Marx. By studying rapidly growing population of the world, Malthus stated that since over population increasing at a faster rate than the food supply, the world will face the problem of over population, this, Marx argued that a particular set-up of the society created the problem of over population. During the 19th century, due to improvements in medical science, the death rate started declining rapidly. This led to an increase in the expectancy of life. Old-age pension schemes were introduced at many places. All this resulted in studies related to mortality. Studies related to fertility began at the end of the 19th century. T.H.E. Stevents analyzed the data related to fertility provided by the 1911 census in England. Dieterici tried to estimate the world population in 1859. The scope of demography widened substantially during the 20th century. Systematic studies of the factors affecting changes in population were undertaken. Attempts were made to study the relationship between fertility and mortality on the one hand and the age-sex composition of the population on the other. Growth of population and problems related to it were presented by Carr-Saunders in his book 'The Population Problem: A Study in Evaluation. The United Nations played a major role in the development of this discipline all over the world, particularly encouraging studies related to population in developing countries. The United Nations published several books and booklets related to the population problem. It also provided financial assistance and encouragement towards the development of this discipline. Population geography remained neglected for quite some time. In the early stages, it was not considered an independent branch of geography. This aspect of geography was neglected even by Hartshorne and Dickinson. Dickinson, in his book 'The Making of Geography' while explaining the development of human geography, has not even mentioned population geography as an independent sub-branch. Earlier, Alfred Hettner, a well-known geographer, wrote that population influences geographical factors in its vicinity. He, however, did not go beyond this and did not make any serious effort towards giving population geography the stature of an independent sub-branch. Huntington and Shaw, in their book 'Principles of Human Geography,' have not made any direct reference to population. In short, population geography remained neglected until the Second World War. Following were the reasons for such neglect. Censuses were not conducted in many countries till the Second World War. Population figures for all parts of the world, therefore, were not available. During this period, geography placed stress on regional differentiations only. While studying the regional differentiations, population was not considered as an important factor.

In the earlier periods, geography confined itself to only regional geography. It provided information on various regions and the distribution of geographical factors. The relationship between geographical background, geographical personality, and factors on the one hand and man on the other was not studied. Many underdeveloped countries became independent after the Second World War. Population became a hindrance to the increased aspirations of the people for development in many countries. Studies related to population in relation to geographical background and natural resources were undertaken in many countries. In 1953, G.T. Trewartha gave population geography the status of an independent discipline. On April 2, 1953, Trewartha, in his presidential address to the American geographers, emphasized the importance of population geography. This led to the emergence of population geography as an independent discipline. G.T. Trewartha was the first geographer to frame curriculum and to include population geography as a subject at the undergraduate level. After this, the teaching of population geography spread to many American and other universities around the world. This effort was further substantially helped by Clarke. Because of these efforts, population geography today has the status of an independent social science. This discipline was enriched by several geographers, such as Beaujeu Garnier (1936), George Demko (1970), Harold Rose (1970), Peters, and Larkin (1979), who wrote books throwing light on several aspects of population geography.

The concern about population in ancient societies was restricted to replacement of population lost through the unusually high mortality. However, during the classical Greek period, one finds some definite views on population. Plato, the great Greek philosopher, emphasized that for human perfection, stability of population is indispensable. In other words, he has emphasized the importance of quality over quantity of population. More or less at the same time, one can find writing showing too few people is greater evil than too many. Kautilya, an Indian, was of this view. At the height of the Roman Empire, one finds the pro-natalist policy more common. Cicero distinguished that rapid population growth is essential to replace war fatalities; this was an era of colonization, which would be easier with more people. However, with the decline of the Roman Empire, an anti-natalist policy started gaining ground. It was a period of economic stagnation and the "fatalistic period" of European History. Celibacy was considered to be the highest form of human being. Through the thirteenth century, slight changes in the anti-natalist policy were noticeable. For example, Thomas Aquinas argued that marriage and family building were not inferior to celibacy. With the growth of mercantilism, where nations were built on increasing trade and wealth, population assumed new importance. In this unit, it is expected to learn about various theories of population. These theories have been presented from their historical perspective.

Difference Between Population Geography and Demography

In population geography, characteristics of a region and the people and groups of people who create imbalances in the region will be studied. Imbalances in the region will be studied. Demography word has been introduced by Greek language. Demos mean people, and graphy means writing. Writing about the people means demography. Demography studies data of population, but population geography studies the relationship between population data and region. Demography is based on statistics, but population geography is based on maps. Demography data will be taken on the basis of administrative boundaries. In population geography data has been taken on the basis of geographical regions. In Demography, generally, humans or persons have been studied. Human groups have been studied in population geography. Demography studies components of population like birth, death, migration, marriage, etc. and their regional differences and changes. In population geography, these changes have been studied on the basis of political and biological aspects, and the factors responsible for these changes have also been studied. Study in Demography has limited scope, whereas this study of population geography has vast scope.

Approaches to the Study of Population Geography

The main approaches to the study of population geography are following:

- **Regional approach:** Region has been taken into consideration.
- **Systematic approach:** The study will depend on climate, physical, social, and economic conditions and their effect on population.
- **Behavioral Approach:** In this, as per regional differences, human behavior also changes. There have been studies in this approach.
- **System Approach:** Population and its relation with other aspects like attributes, characteristics, correlation, and environment have been studied.
- **Quantitative approach:** population is dynamic; to study the dynamism of population, quantitative approaches have been used. This approach was started after 1960.

Sources of Population Data

The Indian Census began in 1872, and it is the largest source of statistics on the people of India. It brings out statistics every 10 years. For scholars and researchers, the Indian Census is an interesting data source.

- **Census:** There are the following tables for different types of data.
- **A table:** population-related data for different regions
- **B table:** economy-related data.
- **C Table:** Social and cultural characteristics of the population have been covered.
- **D Tables:** Migration-related data have been available in this table.
- **Census Methods:** Population censuses use one of two methods

De facto, meaning enumeration of individuals as of where they are found on the census day, where they stay for last night, regardless of where they normally reside. De jure, means enumeration of individuals as of where they usually reside, their home, regardless of where they are on census day.

Registration of vital events: births, deaths, fetal deaths, marriage, divorce, and adaptation data will be available in this system. Continuous population registration: boundary change and place change data are available in this source. Sample survey: In between two census years, the data is available through this source.

Major Sample Surveys in India

The main sample survey that occurred in India is as follows:

- National sample survey: Data is available on a yearly basis from 1950. Birth, death, population increase, internal migration, economically active population, jobs and unemployment, urban labor force, consumer expenses, industrial production, literacy, and urbanization are some of the components.
- Sample surveys by the Gokhale political and economic institute: This institute is conducting data under different projects at the district level of different states of India. The data is available from 1953 onwards.
- National family health survey: The International Institute of Population Studies has conducted surveys related to the health of women and children at the district level of India. There are three phases of this survey: i. 1993-94, ii. 1998-99, and iii. 2005-2006. iv.2015-2016 v. 2019-2020. Recently, sixth phase have been started.

Other sources: Other than the above sources, like data collected by government offices, educational institutions, private institutions, and non-government organizations (NGO). Bruisers have collected the data in the pre-independence period, like gazetteers, which is relevant and useful till data.

Population Theories

The dynamic nature of population makes scholars and thinkers curious about the population question. The points of view have differed, especially as related to the size and growth of human populations. A large and rapidly growing population has sometimes been considered desirable as a source of the nation's strength and wealth and as an essentially useful factor underlying technological development. At the other extreme, it has been viewed as a contributing factor in poverty and such devastations as wars, famines, and epidemics. The size and growth of the population have thus been viewed as an important factor underlying the development of the country. It is, therefore, worthwhile to trace the development in the thinking of scholars and the different points of view expressed by them with respect to different phenomena. In ancient times, several statesmen and thinkers applied their thought to the question of the desirable size, need, and growth of population for either encouraging or discouraging. In the true sense of the term, a population theory can be considered to have emerged only in the eighteen centuries when the well-known work of Thomas Malthus was published, though some thought was given to population issues in the prior period. In this chapter, it is therefore proposed to discuss the various theories as they relate to pre- and post-Malthusian periods. Early thinking on population issues Confucius and other Chinese thinkers: The great Chinese philosopher, Confucius, and scholars belonging to his school of thought, and other Chinese thinkers, had given the concept of optimum population and its relation to agricultural land.

As per this thought, marriage, family, and proliferation were generally positive for population increase. Greek Thinkers: The population theories and policies of the Greeks and Romans may best be understood with reference to their ideals. In the Greek system of political life, the individual was only a part of the state and had to play a secondary role to it. In ancient Sparta, marriage was considered an institution formed by the legal and political system to provide the state with inhabitants and citizens. The real purpose of marriage was the procreation of children. Continuous wars, which decreased the population size, demanded a constant supply of young men. All Spartans were consequently compelled to get married. Celibacy was punishable by law and criticized in public. The state required that all marriages should be good, in the logic that children complete in body and mind should be born of such marriages. Fathers, who had three or four sons, were publicly rewarded in Sparta. The people from Sparta, however, were also very concerned about the quality of their population and disfavored the overburdening of the state with useless inhabitants. As a result, deformed babies were consistently removed. In Athens, the rules concerning procreation were somewhat less rigid, though Athenian customs and laws also encouraged frequent childbearing. The Athenians were also concerned with sustaining and improving the quality of the population. From this, it is obvious that the Greeks were concerned about the size of the population more from the points of view of defenses, security, and government than from that of economic resources. Plato (427-347 B.C.) was more specific on this point when he stated that if the "highest good" was to be achieved, the city-state should have 5,040 citizens, and land, houses, and property should be equally distributed among them. It may be noted that Plato referred only to citizens and did not include women, children, and slaves in the ideal number of 5,040. The total population of such an ideal city state was about 50,000. He recommended that women should get married between the ages of 16 and 20, and marriage should be made compulsory for girls between these ages. Jowett is of the view that there was some element of numerology in the choice of this particular number. This, however, does not explain the choice of the particular number 5,040, for a half of this number of its double would have served that same numerological purpose. Hutchinson says that the purpose of this choice of figure in Plato's planned state was simply to aid in maintaining an equal division of property." Another plausible explanation is that Plato wanted the city-state to be large enough to ensure economic self-sufficiency and military defense but small enough to permit a constitutional government, with substantial interaction among citizens and scope for their straight involvement in the affairs of the state. Aristotle (384–322 B.C.) was of the view that an excessive number of people would give rise to poverty and other social ills since it was not possible to increase land as rapidly as the size of the population. As preventive checks, he suggested abortion and property as rapidly as the size of the population. To sum up, the Greeks were interested in maintaining population size, which was appropriate for a city-state. They also highlighted the quality of the population. They were interested in political, rather than economic, and their politics were the politics of small city-states. Romance Thoughts: The Romans views are regarding conquest, power, and empire. They, therefore, needed an expanding population. From their writings, it appears that they encouraged procreation by grating certain privileges to married couples with procreation by granting certain privileges to married couples with children on the hand and by taxing and disowning those who did not get married or have children.

Religion and Population Issues

Hinduism: Conventionally, the Hindu religion has laid pressure on marriage and the proliferation. Even today, marriage is universal among the Hindus because it is anticipated that a man should go through the numerous stages of his life performing the duties, and marriage is considered to be one such duty. After the phase, *Bramhacharyashram*, when he is a student and devotes his time and energies to the pursuit of knowledge, a man arrives in the second phase, *Grihasthsshram,* by getting married. The aim of Hindu marriage is said to be dharma, *praja* (progeny), and *rati* (pleasure). When a man gets married and becomes a householder, he accepts dharma as the highest aim of marriage, demonstrating that it may be desired not so much for sex or offspring as for finding a spouse for the fulfillment of one's religious deities. For the women, marriage is essential because a man goes through several ceremonies throughout his life; marriage is the only ceremony that she is allowed. Reproduction is the second object of Hindu marriage. Kautilya, well known as the author of Artha Shastra, highlighted the desirability of a huge population for military and economic power.

Judaism: The scared books of the Jews highlighted the need for procreation. The oft-quoted injunction to Adam and Eve by the Almighty is evidence of this: "Be fruitful and multiply, and replenish the earth." Childlessness was considered a grave misfortune. A large population was to the pro of the Jewish people.

Christianity: Christianity was mainly a moral and religious force, and therefore early and medieval Christian writers viewed the question of population purely from the moral and ethical points of view. There was heavy emphasis on celibate practices, the foundations of which are found in the teachings of St. Paul and St. Augustin. Marriage and reproduction were regarded as required ills. In later years, however, this emphasis underwent a change: marriage was said to be desirable, and reproduction was thought of as a divinely approved mode of living.

Islam: The Muslim thinking on population was mainly populationistic. Of the many objectives of marriage, one of the supreme importance is man's procreation of sons and daughters, who thereby renew and extend his own life. Islamic texts are also clear on the point that the primary objective of the establishment of a family is the perpetuation of the human species through the fulfillment of the sexual desire, which is achieved through sexual intercourse between husband and wife. It is, however, worth nothing that a significant work of an Arab author of the fourteen centuries, Ibn Khaldon (1332-1460), made two interesting points. He was of the opinion that a densely settled population was favorable to higher standards of living, for it helped achieve a greater division of labor and a more active utilization of resources and also guaranteed military and political safety.

Pre-Malthusian Theories of Population: Beginning of Sixteenth Century to End of Eighteenth Century: This period is known for the rapid changes that took place in some facets of human life, the most prominent of which were due to the Renaissance or the restoration of art and letters under the impact of classical models (fourteenth century to sixteenth century). During this period, new routes to India were discovered, America was exposed, gun power and printing were devised, and a new stimulus was provided to the arts and sciences as well as to trade and manufacturing. The old feudal structure collapsed, making way for capitalism. This period was also notable for the rise of such powerful states as England, France, Spain, and Portugal. It was ingestible that all these should have some influence on the thinking on economy and population.

The Mercantilist School: In the history of economic thought, "Mercantilism" is considered to be an association between the medieval period and the modern period. Almost all the trading nations adopted this economic policy of mercantilism with a view to growing national wealth and power by inspiring exports of goods in return for gold. The general approach was that the birth rates should be increased for the purpose of economic and political gains by the adoption of such measures as (a) placing various disabilities on celibates; (b) encouraging marriages; (c) encouraging fertility; (d) making punishment for illegitimate births less severe or demolishing such punishment entirely; and (e) encouraging immigration and controlling emigration. Among the early mercantile writers, the most important are Niccolo Macciavelli (1469-1527) and Giovanni Botero (1540-1617). Macchiavelli was perhaps the first to view population from the modern angle, observing that excessive population would diminish through want and disease. In the sense that he may be considered one of the precursors of Malthus. Giovanni Botero, an Italian citizen, was one of the first to study population phenomena in a scientific manner. Along with Macchiavelli, he may be considered one of the precursors of Malthus, as his work contained the basic thoughts of the Malthusian doctrines. He expounded that a population, after increasing for some time, cannot continue to increase at the same rate if it may grow slowly or may even start declining. In his opinion, the limitation of the means of sustainable development was the reason why population growth had to be limited. He regarded limited means of subsistence as the primary check on population growth, while the secondary checks were sterility of the soil, bad climate, disease, and epidemics. Through Botero was convinced that food for subsistence was necessary for any population, he still regarded a large population as a source of strength for the country. The physiocratic school of economics, which evolved in France about the middle of the eighteenth century, may be considered the expression of a reaction against the mercantilist ideas and politics. While the mercantilists had neglected agriculture in their enthusiasm for the acquisition of gold and silver through trade in manufactured goods, the basic assumption of the physiocratic school was that land was the source of all wealth, and hence it was necessary to emphasize the importance of agriculture and internal tax reforms. As a reaction against the population tenets of the mercantilists, the physiocrat did not favor population increase at the cost of standards of living. They approved of such an increase only if it was possible to expand agricultural production. Similarly, views were expressed by Mercier de la Riviere, through whom he maintained that in a well-administered state, population would not increase faster than the means of subsistence. The Beginnings of the Malthusian Theory: from the middle of the eighteenth century, almost all writing on population contained some reference to the point that population increases more rapidly than food supply, a point which was later to be elaborated by Malthus and which came to be known as the Malthusians theory of population. Malthus In 1798, Mathus, then a curate of small Surry Parish, published his essay on population, which was mainly directed against the optimistic views of William Godwin, an English writer and philosopher, and against marinc Jcans *Antoine Condorcet*, a fresh mathematician, economist, and philosopher. The title of the essay was An Essay on the Principle of Population as It Affects the Future Improvement of Society, With Remarks on the Speculations of Mr. Godwin, M. Clandorcet, and Others. This work was published anonymously, for it was feared that some oversensitive persons might be shocked that a person had written a book on population. This significant work of Malthus is a milestone in the history of population studies and has become one of the most famous and, at the same time, one of the most controversial books of modern times. It brought great honor to its author but also a great deal of abuse and criticism. It is significant that the year of publication of this book is reckoned as a base year for the study of population doctrines. For many years to come, all views on population were classified as per, Malthusian, Malthusian, anti-Malthusian, and neo-Malthusian.

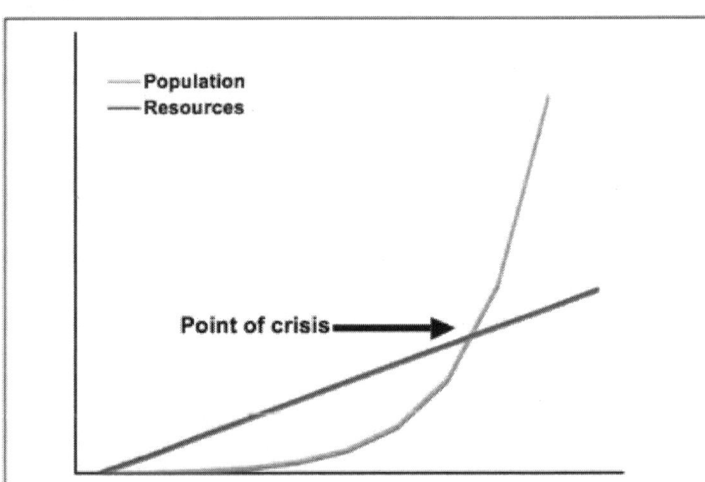

In 1803, Mathus published the second edition of his essay, a much expanded and changed edition that cannot really be called a reprint of the 1797 essay, for in the new edition the emphasis was more on arguments against the poor laws than on counterarguments against the opinions of Condorcet and Godwin. The Central Theme of the Essay by Malthus: Two postulates: 1. Food is necessary to the existence of man. 2. The passion between the sexes is a result of procreation. Once accepted these two laws, population, when unchecked, increases in geometrical ratio. 1. Population tends to double itself every twenty-five years, thus increasing in a geometrical ratio. Human species would increase as the numbers 1, 2, 4, 8, 16, 32, 64, 128,256 2. Even under the most favorable conditions, agricultural produce increases each twenty-five years only by an equal quantity, thus increasing only in an arithmetical ratio. It means and the subsistence as number 1, 2.3,4,5,6,7,8,9.

The Positive Checks: In positive checks, insufficient food and clothing arising from poverty-bad nursing of children—excesses of all kinds will increase large towns and Industries. Mathus classified these positive checks into two categories, the first were brought about by natural causes like common diseases, epidemics, infanticide, plague, and famine, which he labeled as "exclusively misery," The other being that which mankind brought upon itself, such as war excesses of all kinds, which were avoidable but which were brought about by vice and were the consequences of misery. Preventive check on population growth, The postponement of marriage and even permanent abstinence from sex. Couple to limit the number of children according to their wishes. Malthus included moral control and vice as voluntary checks.

Criticism on the Malthusian Theory is: He placed emphasis on the limitation of land. The agricultural revolution of the nineteenth century, which increase in agricultural production. The negative approach of Malthus, therefore, did not come true. The Malthusian theory of population has been severely criticized on several grounds. One point, however, has to be conceded. Despite its several inherent weaknesses, it has to be conceded. Despite its several inherent weaknesses, it has been revived time and again. Though by 1900 his main thesis had been revived time and again, it came into prominence in the second decade of the twentieth century. But this wave, too, subsided by 1934. After the Second World War, however, there was a revival of interest in Malthus. The reasons are, of course, obvious, some of which are: the increased rate of population growth, greater awareness of the consequences of rapid population growth, greater awareness of the consequences of rapid population growth, and the realization that certain natural resources are on the verge of exhaustion.

Optimum Population Theory

An optimal population is a point at which the population is neither too large nor too small but is just enough to secure a maximum return per head under the given conditions of production (Lionel Robbins). Thus, the optimum is determined by (1) the available natural resources; (2) the skill knowledge and habits of the population; and (3) capital and technological progress. Given these factors, there is one "right" population, and that is the optimum. The basic assumption limiting the theory is that other things remain constant. The relationship between population and natural resources are basis for the optimum population theory. Edwin Cannan (1861–1935), define the concept of "optimum population." The first German professor Karl Winkelblench (1810-1865), who, while describing population theory and policy, classified nations into three categories according to the size of their population: Under-populated nations Over-populated nations Nations with normal populations, meaning a size favorable to the greatest possible productivity. Cannan used the term "optimum population" and explained that at any given time, the population that can exist on a given land, consistent with the greatest productiveness.

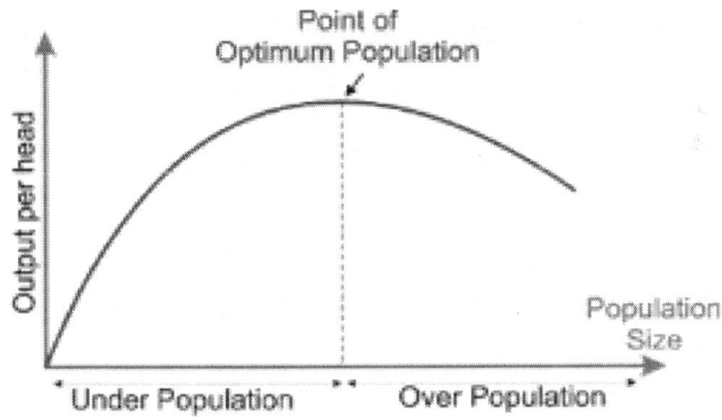

The concept of optimum population has been interpreted in several ways: "to means the size of the population which results in the highest per capita income, the highest productivity as measured in different manners, or the highest level of other less well-defined economic indicators, such as economic welfare, level of living, real income, and, in some cases, employment." Some writers considering the concept of the economic optimum as being too restrictive have included in it the total well-being, health, longevity of a nation, the ideal family size, the conservation of natural resources, power, defense, and other spiritual, cultural, and aesthetic factors. According to most writers, however, the economic optimum was the main consideration in the optimum population theory, and gradually the idea of a population of optimum size for maximum production was accepted.

Marx and Engels: Marx postulated that there could be no natural or universal law of population growth in this view, the population laws are peculiar only to the capitalist system of production, and the tendency of overpopulation was inherent in that system and was not due to excessive reproduction. Engels, while fully supporting the views of Marx, added one point of his own. He maintained that under the capitalistic system, surplus population was associated with surplus capital and that such a contradiction inherent in capitalism could be overcome only by "social reorganization."

Social Views on Birth Control: The Communist Manifesto of Marx and the Origin of the Family of Engels emphasized the need for the emancipation of women from household drudgery. It was, however, not made very explicit whether they should also be emancipated from excessive childbearing.

Mathematical Theories: The availability of data on population facilitated the verification of those laws. Quetelet, a Belgain astronomer, was the first to adopt a mathematical approach to the study of population. The most important among the mathematical theories was the "theory of logistic" growth of population developed independently by Verhulst and Reed's logistic law. The growth of population occurs in cycles and within cycles and in a specially limited area or universe. Growth in the first half of the cycle starts slowly, but the absolute movement per unit of time increases steadily until the midpoint of the cycle is reached. After this point, the increment per unit of time becomes steadily smaller until the cycle. The total population according to this law of the logistic curve follows an S-shaped curve.

The Density Principle: Michael Thomas Sadler, an English political economist and liberal (1730-1835), in his voluminous work entitled. He, however, also made an independent contribution to population theory when he put forward the principle that fertility varies inversely with the density of population. In some countries, like the Netherlands, for instance, both fertility and density were observed to be high. Sadler explained this apparent contradiction by pointing out that a direct relationship existed between fertility and mortality.

The diet Principal Theory: In 1841, another English political economist, Thomas Doubleday (1790-1870), attempted to establish a relationship between population growth and the diet of the people. According to Doubleday, poverty stimulates population growth, as the diet of the poor is inadequate. In 1952, Jouse De Castro, in his book, Geography of Hunger, contended that those who are well-nourished. He based the content of the diet on the birth rates of various countries. Biological Theories of Spencer and Gini: in 1880, Herbert Spencer postulated that fertility increases in response to the progress of society and that in fact is a great agency for mankind's advancement. Another biological theory was put forward by an Italian biologist, Carrado Gini, in recent times. This theory, known as the cyclical theory, says that "population tends to follow an evolution that is similar to that of the life cycle of the individual, passing through the successive stages of development, maturation, and involution.".

The Theory of Demographic Transition

The theory of demographic transition is based on the real demographic experience of Western countries, which have moved from high mortality and high fertility with consequent slow growth of population to low mortality and low fertility, once again leading to a slow growth of population. Earlier demographers such as Landry and Warren Thompson had attempted to construct a typology to describe the transition from a condition of high mortality and high fertility to a condition of low mortality and low fertility. In 1947, C.P. Blacker attempted to identify the following five phases of the demographic transition:

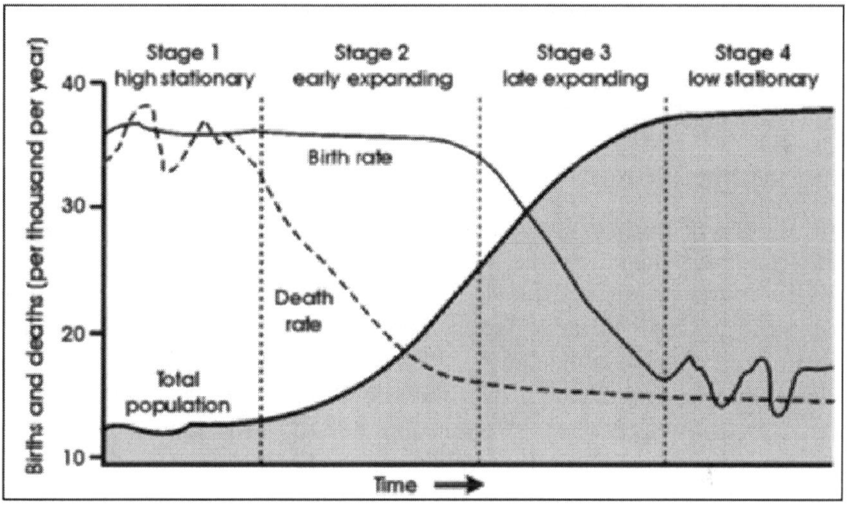

The high stationary stage, considered by high birth rates and high death rates; The early expanding stage, with falling birth rates but rapidly decreasing mortality; The late expanding stage, with falling birth rates but rapidly decreasing mortality; The low stationary stages, with low birth rates balanced by equally low mortality; and the declining stages, with low mortality and deaths exceeding births. None of these demographers can, however, be called the fathers of the theory of demographic transition, for none tried to explain these changes either in fertility or mortality. In 1945, it was Frank W. Notestein who presented the theory of demographic transition in an almost mature form. That Senan explanation is credited with expounding the theory of demographic transition., with an explanation for the changes in fertility. In that sense, he may be credited with expounding the theory of demographic transition. Frank W. Notestein pointed out the rapid growth of population during the past three centuries was mainly due to the decline in the death rate, resulting from the process of modernization, which involved rising standards of living, rising incomes, and advances in sanitation and medical knowledge. Fertility also registered a decline; this response to moderation was not as spectacular. Throughout the modern West, birth rates reached very low levels by the middle of the 1930s. This decline was achieved because of the widespread acceptance of contraception under the influence of the new ideal of the small family, so common in any urban industrialized society. Frank W. Notestein characterized three types of populations according to their stage of demographic evolution. Population in the stage of "incipient decline," where fertility had declined below the replacement levels or those approaching this stage (e.g., population of Europe, the United States, Australia, and New Zealand). Population in the stage of "transitional growth," where "birth and death rates are still high and growth is rapid, but the decline of the birth rate is well established." (e.g., population of the Soviet Union, Japan, and some countries in Latin America.) Populations in the stages of "high growth potential," where "mortality is high and variable and is the chief determent of a downward trend. In these populations, rapid growth is to be expected just as soon as technical advancement make possible a decline in mortality (e.g., the population is most countries in Asia, Africa, and Latin America).

Demographic Transition Theory and Criticism

The theory of demographic transition is accepted widely. It is to be noted that this theory is based on the actual experience of the changes in vital rates in Western countries during the various stages of their industrial and economic developments. Recent studies expose that, in Spain and in some countries of Eastern and Southern Europe, fertility decline experienced even when mortality was very high. In some countries, like the United States, the growth rate in the post-transition stages was probably higher than in Stage II and Stage III of the demographic transition. Notestein's claim that fertility declines initially in urban areas is found to be false for some countries. For instance, countries with predominantly rural populations, such as France, Sweden, Finland, and Bulgaria, did experience declines in birth rates to the same extent as did some highly urbanized countries, such as England and Wales. Such exceptions suggest that the theory of demographic transition is only a broad generalization, which does not experience by all western countries. This theory cannot explain the phenomenon of "the baby boom" in western countries after the economic recovery and the Second World War. It does not give explanation of decline of fertility in Western countries., Does the theory of demographic transition apply to developing countries? It is well known that developing countries have experienced a phenomenal reduction in death rates and recently experienced a decline in birth rates as a result of more population. In the past, there has been a tremendous increase in the rates of population growth. It is applied for developing countries, but the timeline may be different from developed countries.

CHAPTER-2

Population Composition

Population composition refers the distribution within a group of people of specified individual attributes such as sex, age, marital status, education, occupation, and relationship to the head of household. The age-sex composition of a society's population has significant implications for the current and future development of that society. The age-sex structure can cause a population to grow even when the levels of fertility and mortality suggest equilibrium or even decline. In addition, the age-sex composition of a population has its own direct implications for public policy. A Global View of Population Composition in 2002 Covered Some Striking Regional Differences The age-sex composition of the globe's population in 2002 is depicted in following graph

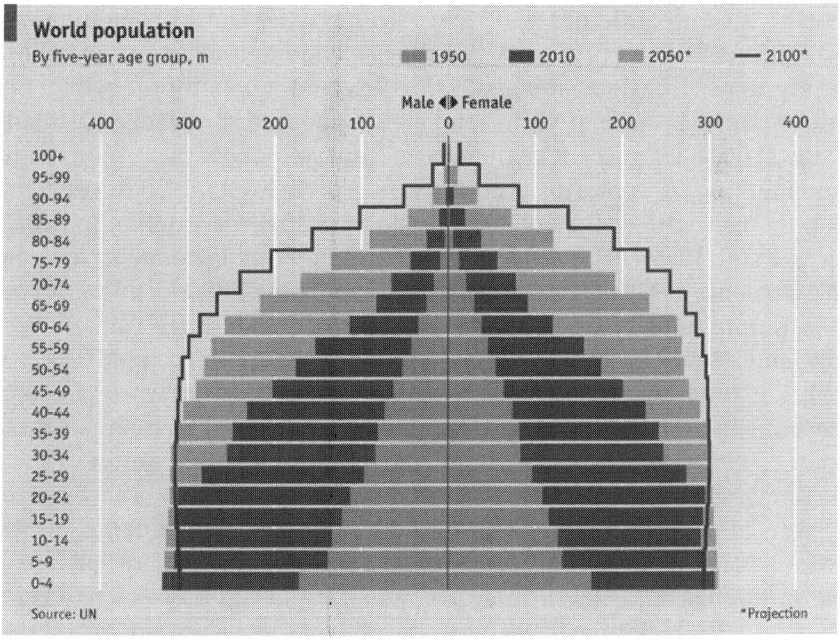

This pyramid illustrates the following characteristics:

The number of men and women on the globe was approximately equal. Men outnumbered women up through the 50-54 age groups, after which women outnumbered men. The largest age group for both males and females was ages 10-14, followed closely by ages 0-4, and then ages 5-9. Each successively older age group, beginning with ages 15–19, was progressively smaller. As the pyramids in Figure 19 illustrate, however, population age-sex structure varied across major world regions. First, although men outnumbered women in all countries of Asia, the Middle East, and North Africa, women outnumbered men in the developed world, Eastern Europe and the New Independent States (NIS), and Latin America and the Caribbean. In addition, although women typically outnumbered men in the older age groups, the exact age at which this transition occurred varied widely across regions, from a low in Latin America and the Caribbean (the 30-34-year-old age group) to the extreme in India, where women never outnumbered men. In relation to the age structure of populations, regions covering less developed countries tended to have the highest percentages of their populations under the age of fifteen. In Sub-Saharan Africa, the Near East, and North Africa, and for Asia and the Pacific, excluding China and India, the largest proportion of the population was 0 through 4 years old. In contrast, in more developed world regions and China, older ages composed the largest population groups. Finally, whereas the populations of less developed regions decreased in size progressively as age increased, the patterns in the developed world, Eastern Europe and the NIS, and China were more irregular and in much larger proportions. The world's population pyramid is changing shape. The world's population will have reached 8 billion by the end of October 2024, according to the latest projections from the United Nations. For the first time, the UN has tried to look as far ahead as 2100, using several assumptions about how fertility and mortality rates might change over the years. It is estimated that the global population will cross 10 billion by 2085. By 2100, 22.3 percent of people will be aged 65 or over, up from just 7.6 percent in 2010. The majority of population growth is expected to come from the developing world. Africa's population will increase from 1 billion in 2010 to 3.6 billion in 2100. In 1950, the world's population living in today's rich countries was 32 percent. By 2100, the population will have declined up to 13 percent. Urbanization in India 2011 The pace of urbanization has been steadily growing in India, and 31.35 percent of India's populations were inhabited in urban areas in 2011. The phenomenal increase in census towns during the 2011 census was also responsible for the increase in urbanization. There has been an increasing tendency to create new districts in different states of India. Amongst the major states, Tamil Nadu, Maharashtra, Gujarat, and Karnataka are more urbanized. The rapid internal migration from rural to urban is also responsible for noteworthy transformations in population densities between rural and urban areas. The decline in population growth rate was moderate in the poverty-stricken states of Bihar and Uttar Pradesh. Several states have experienced deceleration in population growth during the last decade compared to the 1990s. Kerala has registered an impressive annual growth of 0.48 percent. It is alarming to note that there was acceleration of population growth in the states of Tamil Nadu and

Chhattisgarh. The share of small villages in rural populations has been shrinking while that of villages with populations above 5000 has been increasing. Nearly 56 percent of districts have a population ranging from one to two million. Rural-urban disparities on vital rates, though narrowing down over the years, similar interstate differences continue to exist. Expectancy of life has been rising, on average, at the rate of half a year per annum since Independence. In this regard, among the major states, Assam, Madhya Pradesh, and Odisha have been lagging behind other states. The female has better situation in this indicator. It is observed that the sex ratio is declining and the recent improvements on this indicator reveal that Indian society is sensitive to the gender disparities. The sex-ratio has increased considerably during the last decade in urban places of India. As a result of urban growth, the absolute rise in population size and declining average household size suggests the disintegration of the joint family system in favor of the nuclear family. The current crude birth rates, infant mortality rates, and child malnutrition point out that India has to invest heavily in human development to realize the Millennium Development Goals (MDGs).

Population Parameter for India 2011

Population growth

According to the Census India in 2011, the total population of the country was 1,210,193,422. In 2001 it was 1,028,737,436, it increased by 181 million people in one decade. Presently, the most populated states of India are Bihar, West Bengal, Uttar Pradesh, and Kerala. The country covers just 2.4 percent of the total earth's surface but is responsible for 17.5 percent of the total population on earth. The growth rate of population in the last decade is 17.67 percent. In rural areas the population growth rate is 12.3 percent, and in urban areas the growth rate is 31.8 percent. Meghalaya has recorded the highest growth rate in rural population, and Daman and Diu had the highest decadal growth rate in urban population during 2001-2011.

Density of population in India 2011 Population is increasing in India; the recent population of India in the year 2014 was 1.27 billion; it is a total 1,270,272,105 people if viewed in numerical terms. India is the second largest among the most populous nations of the world and is just under China, which has a population of 1.35 billion people, or 1,350,044,605. The population in 2001 was 1.02 billion. The population density of India per square km is also increasing. The population density has increased from 324 in 2001 to 382 in 2011 people per square kilometer, which is considerably higher than the average population density of the world in 2011, which is 46 per square kilometer. Data reveal that along with the wide difference with the population density of the world, there are also a lot of differences in the population density of the various states of India. The National Capital Region (Territory) area of Delhi has the highest population density in 2011, which was 9340 per square kilometer. Among the states of

India, Bihar has the highest density, which is 1102 per square kilometer, while Arunachal Pradesh has the lowest population density, 17 per square kilometer. A higher density of population in a region depends on climate, relief, rainfall, humidity, temperature, economic activities, transportation network, population characteristics.

Age Composition in India

Table 1: Age distribution

Age group	Total Percent	Male	Female
0-14 years	31.1	190,075,426	172,799,553
15-64 years	63.6	381,446,079	359,802,209
65 and above	5.3	29,364,920	32,591,030

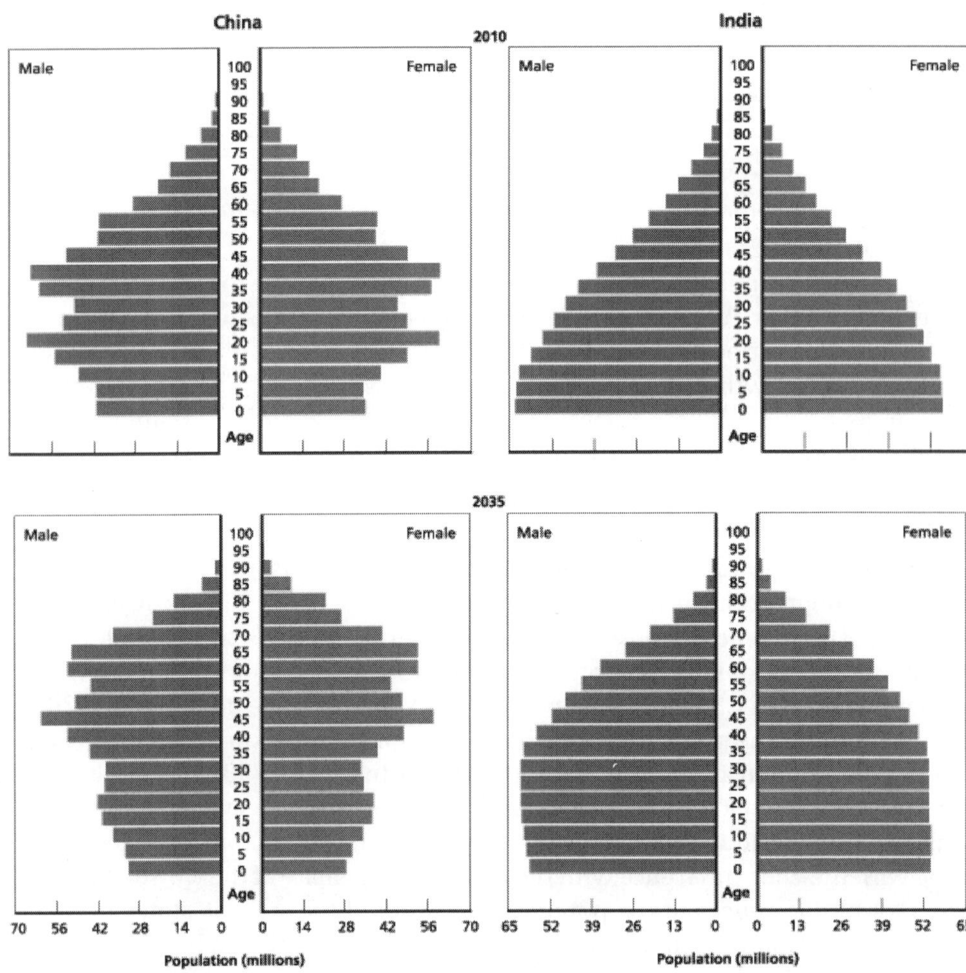

SOURCE: U.S. Census Bureau, International Data Base, 2011.

Sex Ratio of India

The sex ratio is the number of females per 1000 males. As per the 2011 census, the sex ratio in India is 940 females per 1000 males. The sex ratio shows an increase trend in 2011 from the 2001 census data. Census 2001 revealed that there were 933 females per 1000 males. From 1971 the sex ratio started declining till 2001 the decrease in sex ratio has been observed. As per 2011 census, there has been a slight increase in the sex ratio (943/1000). Since, last five decades, the sex ratio has been found around 930 females per 1000 males in India. **(See Map 1)**

Table 2: Sex ratio at different age groups

Age groups	Sex ratio
At birth	1.12 male(s)/female
Under 15	1.10 male(s)/female
15-64 years	1.06 male(s)/female
65 and above	0.90 male(s)/female

The major reason of the decrease of the female birth ratio in India is considered to be the female feticide and brutal killing practices to the girl child at the time of the birth. The sex ratio in India was almost equal during the phase of the years of independence, but thereafter it started declining gradually. Though, the sex ratio of India has gone through commendable signs of improvement in the past 10 years, there are still some states where the sex ratio is still low and is a cause of concern for the Government to take some steps to improve it. Haryana is one of the states that is showing a decreasing trend in the women population in 2011. The state of Haryana has the lowest sex ratio in India, and the figure shows a number of 877 females to that of 1000 males.

There are also states such as Puducherry and Kerala where the number of women is more than the number of men as per 2011 census, in Kerala, the sex ratio is 1084 females per 1000 males. While Puducherry and Kerala are the only two states where the number of females is greater than the number of men, there are also states in India like Karnataka, Andhra Pradesh, and Maharashtra where the sex ratio in 2011 is showing considerable signs of improvement. Some facts related to the sex ratio in India follow: the main cause of the decline of the sex ratio in India is due to the biased attitude that is meted out to the women. The main cause of this gender bias is lack of education and crime against women. Pondicherry and Kerala house the maximum number of females, while the regions of Daman and Diu and Haryana have the sex ratio unfavorable to the female population.

Caste Composition in India 2011

Table 3: Caste composition in India 2011

Caste	Number of People	Percentage
Scheduled Castes	166,635,700	16.2
Scheduled Tribes	84,326,240	8.2

In India, the state with the highest proportion of Scheduled Tribes is Madhya Pradesh (14.7 percent). The state with the lowest proportion of Scheduled Tribes is Goa (0.01 percent). State with the highest proportion of Scheduled Castes, Punjab (20.5 percent) State with the lowest proportion of Scheduled Castes, Goa (0.01 percent). **(See Map 2)**

Literacy in India

The literacy rate of a place is calculated based on the fact that how many total inhabitants are literate. A person is regarded as literate if he or she is within the age group of 7 years or older and is capable of reading, writing, and speaking in a particular language. As per Census of India 2011, the country's literacy rate is calculated in two different ways: Crude literacy rate and Effective literacy rate the literacy rate calculated by taking the total population as a denominator and the number of literates as a nominator is known as the literacy rate. Whereas, while calculating the effective literacy rate, the population aging between 7 years and above is considered. Within the years 2001 to 2011, the effective literacy rate in India increased to 74.04 percent. Out of this, the percentage of male effective literacy rate is 82.14 percent and female effective literacy rate is 65.46 percent. **(See Map 3)**

Table 4: Literacy in India as per 2011 census

Category	Literates	Percent
Total	77,84,54,120	74.04 percent
Male	33,42,50,358	65.46 percent
Female	44, 42, 03,762	82.14 percent

The fifteenth official census in India was conducted in the year 2011. In a country like India, literacy is the main factor for social and economic growth. When the British rule ended in India in the year of 1947, the literacy rate was only 12 percent. Over the years, India has developed socially, economically, and globally. After the 2011 census, the literacy rate in India has increased to 74.04 percent. As compared to the adult literacy rate in India, the youth literacy rate is about 9 percent higher. Female literacy rate is 65.46 percent according to the 2011 census, while the male literacy rate is more than 80 percent. Though this seems like a very great achievement, it is still a matter of concern that so many people in India cannot even read and write. Among the Indian states, Kerala has the highest literacy rate, which is 93.91 percent, followed by Mizoram, 91.58

percent. Among the Union Territories, Lakshadweep having highest literacy rate, 92.28 percent. Bihar has the lowest literacy rate in India with 63.82 percent. The male literacy is highest in Lakshadweep (96.11 percent) and Kerala (96.02 percent). The female literacy is highest in Kerala (91.98 percent) and Mizoram (89.40 percent). The lowest male literacy is in Bihar, 73.39 percent. The lowest female literacy is observed in Rajasthan, 52.66 percent. It shows the disparity among the states and between the sexes.

Workers in India as per census 2011, the total number of workers (who worked for at least one day during the reference year) in India was 48.17 crore. Out of which 33.19 crores of workers are male and 14.99 crores are female. The workers have registered a growth of 19.8 percent. 34.86 crore workers are in the rural areas, and 133.1 million are in the urban areas. The work participation rate (WPR) for the country worked out to 39.8 percent. This is marginally higher than the corresponding WPR of 39.1 percent on census 2001. The WPR for males increased to 53.3 percent in 2011 in comparison to 51.7 percent in census 2001.

Millennium Development Goals (MDGs)

The United Nations Millennium Development Goals are eight that all 191 UN member states have agreed to attempt to achieve by the year 2015. The United Nations Millennium Declaration, signed in September 2000, commits world leaders to reduce poverty, hunger, disease, illiteracy, environmental degradation, and discrimination against women. The MDGs are derived from this declaration, and all have specific targets and indicators.

CHAPTER-3

Fertility

Introduction

Fertility: Fertility is actual birth performance of a woman. In some cases, a woman may have capacity to have children but may not be sexually active. She may not marry. Socio-economic situation may prevent her from marrying after physical maturity. Alternatively, a woman may be married but suffer from sterility. During the adolescent ages (in teens), girls' capacity to reproduce is less even though they can have fully satisfied sexual activities. This is due to adolescent sterility. Moreover, a woman may be fecund and sexually active and still not produce a child because she or her sexual partner is using some contraceptive method to prevent conception.

Basic Concepts in Fertility Studies

Fecundity: Fecundity is defined as the biological capacity of a woman to reproduce or to conceive. The capacity to reproduce remains with women till the age of 45-49 years.

Fecundability: Like their fellow professionals in other areas, medical scientists and demographers would like to measure the phenomena that they study. Fecundability is a measure of fecundity. It may be defined as a measure of chance of conception.

Natality: The analysis of natality refers to certain measures of fertility.

Family Size: the number of children a woman or a couple has at a point in time. The completed family size indicates the total number of children born by a woman up to the end of her reproductive life, which in normal cases ends at around the age of 45-49 years.

Birth Order and Parity: the total births at a particular point in time in the population are classified according to their occurrence, the rank or order of the births. When women are classified according to the number of children born to them, they refer to their parity.

Fertility and Reproduction: Fertility analysis deals with only the child-bearing activity of a population, whereas reproduction refers to the ability of a population to replace itself and to grow.

Sterility: the lack of capacity of a woman or a couple to produce a live birth is called infecundity or sterility. If a woman has never been able to bear children, such woman is called a primary sterile woman.

Secondary Sterility: on the other hand, arises when a woman has already produced a child but has become sterile before reaching the end of the reproductive span or menopausal status.

Reproductive Period: The reproductive period or the child-bearing period begin with the onset of menarche and continues till the menopause. Usually, the age group 15-49 is called the reproductive period.

Amenorrhea: Temporary absence of menstruation (normal or pathological) is called amenorrhea. The amenorrhea period following a birth is called post-partum amenorrhea.

Natural and Controlled Fertility: As mentioned earlier, Natural fertility refers to the fertility which exists in the absence of deliberate birth control while controlled fertility is the fertility which occurs in the context of a deliberate use of birth control.

Sources of Data

Vital Registration System: Among all the sources of fertility data, vital registration is the main source of direct data on fertility. The statutory registration of births and deaths was introduced in India way back in the last quarter of the nineteenth century. In pre-independence India, it was the only source of data on vital rates, *i.e.,* rates of fertility and mortality. Of late, several legal and administrative steps have been taken to ensure the completeness of these data. In this regard, the Registration of Births and Deaths Act was passed in 1969. Yet it is not a dependable source. Registration data are far from being complete. Overall, registration is more complete in urban areas than in rural areas. For certain large cities, such as Mumbai, it is more or less complete, but in rural areas where close to 75 percent of the of the population lives, records are highly erroneous. There is a need to strengthen this database more, as this alone can give direct data on vital rates and their short-term changes.

Population Census: In India, the population census is conducted uninterruptedly at decadal intervals, beginning in 1871, and has been the main source of population data. As a matter of fact, up until the 1950s, the census was the only source of reliable data. Census yielded information on size of population, age, sex, and marital status of all Indians and population distribution according to urban-rural residence, industrial and occupational categories, mother tongue, religion, and literacy.

Sample Surveys: Surveys are now playing a great role in generating fertility data for comparative purposes as well as for studying trends. Surveys aim at collecting specific information on fertility from a representative sample by using scientific designs and trained investigators. Since surveys do not include the whole population, their estimates may differ from true population characteristics.

Sample Registration Scheme: Realizing that the achievement of a reasonably efficient system of vital registration is a slow and long-drawn-out process, many countries, such as Turkey, Liberia, Columbia, Morocco, the Philippines, and Kenya have introduced the Dual Record System, which combines the registration and survey methods. The Registrar General of India introduced the Sample Registration Scheme (SRS) in 1964–65. Today it is recognized as the primary source of reliable fertility (and mortality) data by both national and international agencies.

Theories of fertility Social Capillarity Theory: Dumont (1890), a French philosopher, said that one has an urge to rise and improve his social status. He compared the urge to rise in social scale to the inevitable physical law of nature the force capillarity. It is known that under this force, liquid rises in a very thin tube, oil rises in wick, and sap rises in a tree during the spring. The liquid in a narrow tube has an advantage over the liquid in a wide tube in the sense that it can rise higher. Similarly, in the matter of rising in the social scale, a person having the least family burden that is, a person with a small family will rise faster in the social scale. A force of social capillarity according to Dumont, affects fertility. Column of liquid has to be thin order to rise under the force of capillarity. So, a family must be small in order to rise in the social scale. This force of social capillarity, according to Dumont, affects fertility. Dumont further explained that it is difficult to ride in the social scale in societies where status and caste are rigid barriers to individual social rising. In such societies, fertility is found to be high because of the deprivation of individual progress. In a society where there exists political, economic, and social equality, the social capillarity will be strong, and the individuals will believe that it is their right to advance themselves and hence reduce their fertility. Also, it is not merely the concern for one's own status; concern for children's mobility has also proved to be a great factor in fertility decline in the West, at least in the early phases of fertility decline. The theory of social capillarity was the first logical attempt to explain the decline in fertility. In fact, the basic idea of his theory later influenced the thinking of Kingsley Davis, whose fertility theory of change and response is based on the theory of social capillarity. While the theory of social capillarity has been praised for its emphasis on the motivational orientation, it has been criticized on the ground that it was not backed by sound statistical proof.

Theory of Diffusion or Cultural Lag

The theory of diffusion explains how the innovation of birth control and the concept of birth control speedily spread over the world. According to the cultural lag theory of fertility differential, in countries where fertility has been declining, attitudes and practices toward diminishing fertility have been adopted first by the better educated, wealthier, and socially more advanced groups of the city population and transmitted over time to intermediate and lower status groups and to rural areas. This theory emphasizes that fertility regulation, especially contraception, has been a recent development and has been recently introduced in human culture. Another assumption of the theory is the time lag and trickle-down in the spreading of the ideas about birth control and its practice. Birth control practices are initiated in the big cities, move to other urban centers after some time lag, and fertility reaches the rural areas. The middle-class population accept this innovation of birth control with priority basis, which then trickles down to other lower classes. In support of the theory of diffusion, examples of time lag in the fertility declines between the urban and rural areas have been cited. Writing on diffusion and cultural lag Sorokin (1959) suggested that the diffusion of culture is a gradual process: the cultural objects do not move directly from the highest strata to the lowest strata; they move downward to the closest social strata in steps, and their circulation is associated negatively with the height and rigidity of social stratification and positively with social mobility. In general, time lag in diffusion depends on: (a) exposure and context. (b) nature of migratory cultural phenomenon; (c) degree of special qualification, training, and skill necessary to apprehend the circulating cultural system; (d) nature of the culture of presentation and diffusion; (e) presence or absence of force. The theory of diffusion in the context of fertility came under heavy attack by many prominent demographers. For example, Carlesson (1966) questioned the general belief that birth control was an innovation. In fact, he maintained that the practice of birth control did not spread as an "innovation" but as an adjustment of old-style ideas to new-style ideas. Geoffery Hawthorn, on the other hand, did not lead to fertility decline during the period 1860–1946. Another important criticism against this theory is that it is more of an explanation of "how" fertility declined than "why" fertility declined in Europe.

Theory of Change and Response: In 1963, Kingsley Davis pronounced the theory of change and response in order to explain the declines in birth rates in the developed countries. It is well known that even before the secular decline in birth rates in the industrialized countries, mortality rates had been declining, as a result of which the rates of natural increases had gone up. People in these countries found that the sustained fall in mortality (their demographic behavior), which was brought about by the higher standard of living and improved public health programs, was preventing them from making full use of the benefits of the growing economy of the time. They started feeling the burden of large families from the real situation and life experiences, and therefore people began to change their reproductive behavior, using all demographic means possible, such as postponing of marriages, rising age at marriage, celibacy, use of contraceptives, abortion, sterilization, and to some extent migration. Davis termed these responses as multiphase responses. It may be recalled that in North-West Europe and Japan, economic development and population growth were almost simultaneous. At this time, the aspiration of the people "to go ached and appear respectable was aroused". It was primarily a fall in mortality that was responsible for making it difficult for them to go forward in life, and therefore, they began to modify their demographic behavior. Thus, the connecting link between the stimulus (sustained drop in mortality and resulting increased natural growth) and the response (curtailment in the rate of population growth by increasing the age at marriage, resorting to contraception to an abortion, and even migrating outward) was the fear of the loss of status through rising prosperity. In the theory of change response, traces of the theory of social capillarity are clearly visible (can you identify these similarities?) Studying Japan and Ireland as a case, Davis tried to substantiate with figures the multiphase aspect of people's response to fall in mortality. David Glass criticized this theory of change and growth on several grounds. On the positive side, Glass pointed out that theory is compressive and sophisticated, and Davis has documented the multiphase responses by giving the actual experience of two countries Ireland and Japan. He finds this approach having substantiated the mechanical response to the falling mortality by resorting to a sociological analysis. On the negative side, part of the theory, he pointed out, cannot be applied to the situation in the developing countries where the majority of the people are living in poverty and population growth is not accompanied by economic development, as in Japan and Ireland, and where the ambition to rise in the social scale is not possible. He always pointed out that historical events do not always fit well with the theory. Some European countries have experienced a decline in fertility before the decline in mortality. He also questioned the prosperity of taking mortality declines as the starting point for the change and response theory. This part of the theory, he pointed out, cannot be applied to the situation in the developing countries where the majority of the people are living in poverty and population growth is not accompanied by economic development, as in Japan and Ireland, and where the ambition to rise in the social scale is not possible.

Leibenstein's theory: Liebenstein (1957), in his famous book, Economic Backwardness and Economic Growth, formulated the utility theory of fertility. According to this theory, people make rough calculations about the number of children they desire based on the utility and disutility of children. These calculations consider the balance between the satisfactory or utilities obtained from an additional child and the "cost," both monetary and psychological, of having an additional child. According to Liebenstein, an additional child involves six utilities and two types of costs. The utilities are: (a) consumption utility; (b) work-income utility; (c) economic risk utility; (d) old age security utility; (e) long-run status maintenance utility; and (f) contribution to extend family utility. From an economic angle, consumption, security, and production utilities are the major utilities. The two types of costs involved are: (a) direct cost of bringing up a child; and (b) indirect cost, i.e., opportunities forgone due to having an additional child, such as the mother's inability to work or inability to raise the standard of living. Leibenstein has argued that several changes occur during the course of economic development that affect utilities and the costs of an additional child. Such as income effect, survival, and occupational distribution. Among the three incomes, effect is the most important. Consumption utility of the children doesn't change with an increase in income, and the work (income) utility of the additional children to parents decreases as income (per capita income) levels increase. This is because at a higher income level, the families want to have quality (highly educated) children, and the expectation of income from children is almost nonexistent. The security utility also declines as a parent's income increases. However, the direct and indirect costs of an additional child are likely to increase with the rising per capita income. This shows why a rise in income should normally be associated with a fall in fertility levels (Leibenstein, 1975).

Becker's Theory

In 1960, Gray Becker extended the economic analysis to a new area, that of reproductive behavior. He considered children as durable consumption goods who provide utility to their parents. Further, he made two more assumptions. One is that the "quality" of a child is determined only by the amount spent on it, and better-quality children can be obtained by spending more on them. The second assumption is that every family has perfect contraceptive knowledge; as a result, they can control fertility and are able to space and limit births as desired. Applying the neo-classical micro-economical model of consumer choice, Becker postulated that parents choose the optimum number of children, which balances the utility that they derive from the children against the budget constraint, reflects the wage rate of labor, and the prices of different goods and services. therefore, the reason deciding family size is purely economic. How parents choose the number of children they want to have depends, to a large extent, on the exogenously determined (over which the consumer has no control) prices and income. Variation in the price and income causes a difference in demand for children for a given household over time or among households at a given time. For example, when the price of a child relative to other goods increases, the rational consumer will demand fewer children. We can similarly work out the effect of an income change on the demand for the children: Assuming children and the standard of 'child quality' to be 'normal' goods, an increase in income will cause families to demand not only more children but also better-quality children. Having thus set out the basic model, Becker derived a testable hypothesis that as the income level of a family increases, the number of children demand would also increase. However, empirical evidence refuted the hypothesis. Rich parents did not have more children than poor parents; in fact, they tended to have fewer children. Becker provided an explanation for this contradiction. According to him, knowledge and access to contraception were not uniform across all families. Poorer families did not know or have access to contraception and were thus forced to have more children. The reason why richer families tend to have fewer children is tied to the quantity-quality trade-off. Becker's model accounts for the fact that social class, to a large extent, determines a person's tastes and lifestyles.

Thus, the richer parents would tend to prefer better quality children. Further, quality is enhanced by an increase in the inputs of time and goods spent on children; obviously, a better-quality child would be one who is better educated, is better nourished, or has access to various opportunities. As richer parents have to spend more than poorer parents to achieve higher-quality children, they tend to reduce the number of children demanded. In other words, they emphasize the quality of children at the expense of quality. Despite its innovation appeal, Beacker's theory was criticized by sociologists on the ground that it was a purely economic analysis that ignored the sociological determinants of reproductive motivation. Critics of economic theory ask: Are babies durable? Applicability of microeconomic models of fertility depends on the answer to this question. A sociologist argues that there is no simple connection between incomes and desire for children. The main thrust of the argument is that, apart from being ignorant of contraceptive methods, poorer families can also desire larger families. Also, many parents may avoid having just one child and may value providing siblings companions: (a) unlikely the purchase of consumer facilities; (b) the features or characteristics of babies cannot be tested in the showroom as one can test a consumer durable; (c) childbearing is an irrevocable process, for a child cannot be returned or exchanged as is possible with inanimate products. Fourthly, Becker assumes that parents have a choice over the quality of children they rear, but in reality, this is not so. For example, both parents and children who reside in the same house hold and enjoy the same standard of living. It will be, therefore, difficult for parents to discriminate against their children. You may note that the main intention of this model was to explain fertility differentials in industrialized societies. As such, it will have little practical value for policymakers in developing countries. Since Becker's pathbreaking work in 1960, the 'new home economics has incorporated new features. The first such modification was the inclusion of women's labor force participation into the model; women's time costs of childbearing were thus recognized. It was assumed that a mother has the option of spending her time producing income or rearing children. An increase in her income sets off two different effects, i.e., an income effect that has a positive effect on her demand for children and a price effect that has a negative effect. Empirical evidence from the industrialized countries confirms that usually the price effect dominates the income effect. Thus, women's labor force participation tends to reduce their demand for children. Other modifications include the incorporation of uncertainty, educational aspirations of parents, and consideration of time along with their money income into the budget constant.

Easterlin's Framework: the synthetic framework of Easterlin is an integration of economic and sociological approaches. Easterlin's theory acknowledges that the couple's reproductive capability is a condition beyond the control of the family as it is constrained by the familial environment, which is not within the decision-making purview of the family. The incorporation of this aspect into the analysis had marked a new approach in the study of reproductive behavior. How the model works is outlined below. There are three basic components to the model: the demand for children (Cd), which is the number of surviving children's parents would want if fertility regulation were costless; the supply of children (Cn), which is the number of surviving children if parents made no deliberate attempt to limit family size; and the cost of fertility regulation (RC), which includes both subjective and economic costs of contraception. The demand for children is conceptualized in a manner similar to Becker's parents having a utility function in which the number of children and other goods consumed by parents in the future. Once again, parents face a budget constraint, and they balance their subjective taste for goods and children against the externally given income and price in a way to maximize their utility (i.e., their satisfaction). Thus, parents choose the number of children given that they are not constrained by biological and cultural factors. Unlike Becker, Easterlin does not use the quantity-quality trade-off argument to explain a parent's decision. The crucial innovation in this model is the introduction of the supply factor (i.e., the potential number of surviving children a couple would have if fertility were not deliberately limited). Natural fertility and child survival precepts determine the number of surviving a couple can have. As you already know, biological and physiological factors affect the period of exposure to intercourse, fecundability, duration of postpartum in fecundability, spontaneous uterine mortality, sterility, and the like. Fecundity impairment due to disease or genetic disorder, malnutrition affecting coital frequency, and the ability to carry a fetus to term are some sequences of how natural fertility is brought down. Similarly, cultural and social factors also work to reduce fertility. These include practices such as abstinence and breastfeeding. Parents consider their family size desire along with their reproductive ability to achieve it. This will then affect their determination to practice concentration. There is no motivation to take contraceptives if the potential supply of children falls short of the couple's desired family size. Parents would then have as many children as they possibly can. In other words, the number of children they have will correspond to their potential supply. If, however, the potential supply exceeds demand, then there is scope for contraception to regulate their fertility, and greater the excess of supply over demand, greater will be the motivation. At this point in the model, Easterlin introduces the regulation cost concept. Sometimes, even though a motivation to use contraceptives exists, couples may not use them. Easterlin argues that the reason behind couples's reluctance is that there are costs attached to regulating fertility. Apart from the monetary costs of acquiring and using contraceptives, there are also subjective costs. Examples of subjective costs could include displeasure and fear of using contraceptives. Thus, ultimately, couples balance their motivation to control fertility with the costs of fertility control. If their regulating costs overwhelm their motivation to limit family size, contraception will not be practiced. As a result, the number of children they have will correspond to their potential supply. On the other hand, if motivation is high and the costs of regulation are low, then couples will practice contraception and they will have exactly the number of children they desire.

As you can see, modernization is represented on the x-axis, and the surviving number of children on the y-axis. In the early stages of modernization, demand for children exceeds supply, and hence his number of surviving children (C) falls below the Cn curve. As modernization occurs, supply exceeds demand, and there is scope for fertility control. However, the motivation to control fertility is low and the costs are high, leading to a situation where the numbers of children continue to correspond to supply. As modernization progresses, the motivation to use contraceptives increases and the cost of regulation decreases; as a result, contraception is practiced, and the number of children born will correspond to the CD curve. The stretch between points h and p on the x-axis symbolizes the time it takes to move from a stage of low contraceptive use to fully contraceptive society. Thus, we have the following conclusion: In the absence of contraception, a couple's observed fertility will correspond to its natural fertility level and depend on factors such as period of exposure to intercourse, fecundability, duration of postpartum in fecundability, fetal wastage, and sterility. However, if contraception is practiced, then this will be an additional factor along with the supply and demand ones in determining fertility. Easterlin has paid great deal of attention to the demand side. To explain the swing in American fertility, Easterlin (1969) forwarded a theory that says that fertility behavior of a family depends on its relative economic status, i.e., economic status relative to its aspirations. His relative economic status was measured by the "relative cohort" size in empirical application of the theory to American data. The idea is that each generation focuses on a certain idea of status during socialization. It aspires for that. It is the relative success or failure to achieve the aspiration that affects their fertility plans and thus fertility. Absolute status is not all that important, as assumed by economic theory earlier.

This model, too, has its shortcomings. It is not a dynamic model, i.e., it assumes that parents make their reproductive decisions at the banging of their reproductive career. The reality is, of course, that these decisions are made and modified and altered as couples progress through their life cycle. These decisions are more sequential in nature, unlike the static modeling. It also ignores factors that are important in certain cultures, such as 'son preference'.

Caldwell Wealth Flows Theory: Caldwell (1976) developed the wealth flows theory of fertility to explain the fertility behavior of the population in developing countries. He defines wealth flow in terms of the flow of money, goods, and services from children to parents and vice versa. According to him, lifetime net intergenerational wealth flow would determine the fertility of the population. In a society, fertility is high if children are economically not beneficial to parents and low if next generation are economically not beneficial to parents. The flow of wealth in all traditional societies has been from the younger generation to parents. children are economic assets to their parents in the flow of wealth in all traditional societies. In such a situation, more children mean more wealth, leading to high fertility. It implies that fertility will remain high as long as this intergenerational flow of wealth is from children to parents. further he explained that, in present societies, large family children, daughters-in-law, grandchildren, etc. contribute a lot by working in the household and on the farm. In those societies, having many children is an advantage. families are nucleated in advanced societies therefore, emotional and economic isolation of is high in the developed societies of the world. Parents think for themselves and their children, avoiding extending family life in general.

U.N. Threshold Hypothesis: There had been a general consensus that the observed relationship between development and fertility would be negative and that when this socioeconomic condition starts improving, the less developed countries would undergo the declining trends in fertility. Yet in the 1950s and 1960s, less developed countries did not exhibit declining fertility levels despite improvements in socio-economic conditions.

The UN has represented the threshold hypothesis to explain this anomaly. According to this hypothesis, the development-fertility relationship is contextual. In the early stages of development, small changes in socio-economic levels would have no bearing on fertility levels. When there is no impact of socio-economic level on fertility. It is only when the socio-economic indicators have already reached a minimum level that further improvement in them will bring about changes in fertility levels. There is some limit level of development indicators of starting of fertility decline.

Measures of Fertility

Crude birth rate (CBR) The crude birth rate is the ratio of the total registered live births in some specified year in a particular area to the total mid-year population of that area multiplied by 1,000. It is computed with formula:

CBR = B/P* K

B represents the total number of live births in a year. P represents the total population in the middle of that year. K is 1000.

General Fertility Rate (GFR):

GFR is defined as the ratio of the total number of live births in some specified year in a particular area to the women between the ages of 15 and 45.

GFR = B/Pi * K B represents the number of live births during a year. Pi represents the mid-year population of women between the ages of 15 and 44 or 49. K is 1,000.

Age Specific Fertility Rates (ASFRs): There is a characteristic age pattern of fertility that is very similar all over the world. This age pattern is best revealed by computing ASFRs. The formula for age-specific fertility rates is:

GMFR = bi/pi*K Bi is the number of live births to mothers of a specified age group in the population during a year. Pi represents the mid-year female population in the same age group. K is 1000.

Total Fertility Rate (TFR): TFR is defined as the sum of the age-specific fertility rates of women by single years of age from 15 to 44 and 49 is expressed per woman, i.e., the sum is divided by 1000.

Differential Fertility

Differential fertility is mainly a characteristic of transitional populations. In the first stage of demographic transition, fertility is high in almost all the regions, ethnic groups/tribes, and classes. In the second stage, when fertility declines, the change starts with the higher classes and gradually diffuses to other classes in the population. This is the stage when differentials become larger and larger. Eventually, in the third stage, fertility will be low in all groups and fertility differentials will be minor; there may remain small variations by racial, ethnic, or religious groups. For example, in the industrialized countries, fertility differentials have not completely vanished, but in all the socio-economic and ethnic groups, fertility has moved towards replacement or below replacement level. This relationship between fertility differentials and demographic transition is so deep that in the 1950s and 1960s, when many less developed countries did not have reliable data on birth and death rates, fertility differentials as revealed by surveys were considered as indicators of onset on demographic transition.

Areal Differentials Urban-rural residence Geographic and administrative regions, such as country, state (province), district, etc. Population characteristic differences -education (with greater emphasis on female education); -age, income and occupation/size of landholding, race/ethnicity, religion, status/caste/tribe, -migrants vs. non-migrants, and Minority group status, Fertility differentials exit in India according to all above variables. Data from censuses, sample registration schemes, and surveys confirm them. Fertility differentials not only indicate that the population has entered the second stage of demographic transition, but they also help in building socio-economic, cultural, and psychological theories of fertility. To answer why urban areas, have lower fertility than rural areas or why minority groups have higher fertility, you require theories or hypotheses of fertility transition that can be tested against empirical data. Fertility differentials also help the planners in identifying lagging groups with respect to family planning, estimating the unmet need for family planning in different groups, and evolving appropriate community or area-specific approaches. Inclusion of fertility diffusion in projection models can lead to better projections of population trends. The National Family Health Survey has produced a lot of data in India to exhibit fertility differentials by: (a) state; (b) groups of backward districts in states. (c) urban-rural residence: (d) literacy and education; (d) religion; and (f) scheduled castes, tribes, and general population. It shows that fertility differentials in India are quite pronounced, but in states such as Kerala, differentials have greatly narrowed down.

Adolescent Fertility

Adolescent fertility refers to fertility in the adolescent ages. In the empirical studies of adolescent fertility, it is operationalized as age-specific fertility rate in the age group below 20 or 15-19 years. There are many reasons why adolescent fertility is of concern to demographers. It has the following consequences: Demographic consequences, Obstetrical consequences, Health and survival consequences for infants, Maternal health and survival. Social consequences, psychological consequences, and Child development consequences. Demographic consequences are very obvious. High adolescent fertility results in high overall fertility. This impact of adolescent fertility can be marked more among developing countries, which have a rather young age distribution of population with a high percentage of population in adolescent ages. Further, births among adolescent mothers have more obstetrical problems and affect infants and mothers. Births at early ages are known to suffer from high infant mortality and are also associated with high maternal mortality. Studies have shown that teenage fertility, especially in the West, is associated with high dropout rates among girls in schools and colleges and affects the chances of their career mobility. It leads to divorce and family disorganization in general, having ill consequences for the psychology of woman and child development.

There are big changes observed throughout the developed world. The replacement rate, stabilized population for developed countries, is 2.1, yet nearly half the world's population has birth rates lower than that. In the U.S. the total fertility rate (TFR) is 2.0, nearly the replacement rate, with Hispanic immigrants leading in birth rates. The U.S. has a larger elderly population but less than many other countries. A 2010 census showed that 31.4 million Americans live alone, which is 27 percent of all households (equal to the percentage of childless couples). Living alone leads to lower fertility. Low fertility rates have been observed in Western European countries, which are below the replacement rate of 2.1. Germany has a 1.4 fertility rate (its total population is 81.9 million, of which 8.2 percent are foreigners). Holland has 1.8 (16.5 million, of which 4.4 percent are foreigners). Belgium has 1.8 (10.8 million, of which 9.8 percent are foreigners). Spain has 1.4 (46.1 million, of which 12.4 percent are foreigners). Italy has 1.4 (60.2 million, of which 7.1 percent are foreigners). Sweden, which provides strong support for parents, has a high TFR of 1.9 (9.4 million, of which 6.4 percent are foreigners), but that's still below the replacement rate. Ireland and the U.K. High TFRs, at 2.1 and 1.9, respectively, found among non-European immigrant parents. During the 21st century, the U.S. started becoming the slow-aging leader of a rapidly aging world. Singapore's experience is also the same as in these countries. In developed countries, birth rates have been steadily declining. The fertility rate of the population in China is the lowest, 1.08 (2011), with the rate for Indians 1.09 and for Malays 1.64. Developed countries around the world experienced a massive baby boom in the 1950s, which then fell dramatically to see the fertility rate fall to below the replacement rate in almost every developed country in the world. Fertility rates have been rising in many developed countries, while developing countries have experienced a faster-than-expected decline in fertility rates.

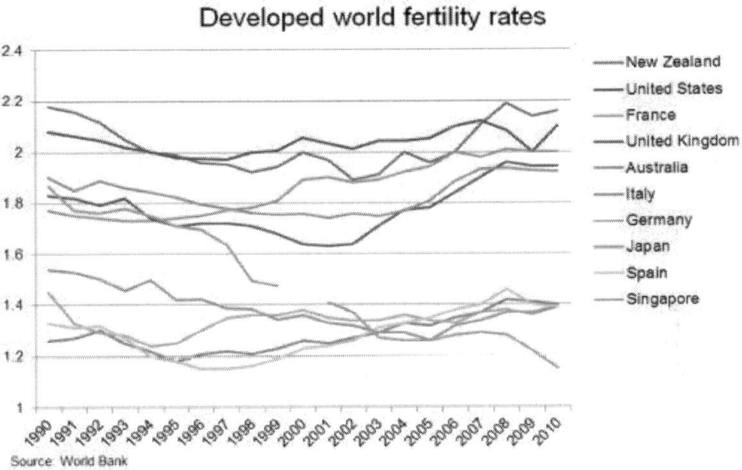

The trend is consistently down in Singapore. The US, which has had the highest developed world fertility due to its growing Latin American, Catholic population, is in fact experiencing a relative 'baby bust'at the moment. However, in the last decade, the fertility rate has been increasing in other developed countries. Spain and Italy in particular, while they have among the lowest fertility rates of Western countries, have risen fairly consistently for the last decade. The United Kingdom has seen fertility rise from 1.65 to 1.9, Australia has had a boom that has now flattened, and France's fertility rate is now almost at the replacement rate. The social rift in the 1960s pushed fertility rates dramatically down as women adjusted to new roles and opportunities. At a certain point, men and women change their priorities, which is economic. The past five years have had an uncertain economic situation, but fertility has not fallen. In the future, the fertility rate may continue to increase. In developed countries today, women receive education and do salaried jobs that are equal to their male counterparts. The fact that women are no longer socially or economically dependent on men and not in household chores. It has radically altered young people's lifestyles. Therefore, the family fabric has been changed; women can now choose to remain single; as per desire, they will choose to have children.

Current fertility rates General Trends: Today developed countries have fertility rates below the replacement level which is 2.1 children per woman.

Countries with notable fertility rates (2021 data): Japan: 1.34 Germany: 1.53 Italy: 1.41 France: 1.83 (one of the highest among developed nations) United

States: 1.78 Average Age of Mothers: The average age of first birth to woman has been increasing: United States: 26.9 years United Kingdom: 30.2 years Sweden: 30.5 years Factors Influencing Fertility Rates Economic Conditions: Economic stability and job security significantly impact family planning. Financial constraints often lead couples to postpone having children.

Work-Life Balance: Work-life balance is a crucial factor. Countries with supportive family policies (e.g., parental leave, subsidized childcare) tend to have higher fertility rates. For example, Sweden and Norway offer extensive parental leave, which supports family growth. Assisted Reproductive Technologies: Increased access to ART has become common, helping couples conceive later in life. For instance, around 2% of births in the U.S. result from IVF. Cultural Shifts: Changing societal norms around family and gender roles also influence fertility rates, with many individuals prioritizing education and careers before starting families. Aging Populations Developed countries are experiencing significant demographic shifts due to aging populations. In countries like Japan and Italy, over 20% of the population is aged 65 or older. This poses challenges for social services and the labor market. Impact of COVID-19 The pandemic affected birth rates, with many countries reporting declines in 2020: United States: Births decreased by about 4% in 2020. Italy: A reported decline of 21,000 births in 2020 compared to the previous year. Conclusion The fertility situation in developed countries is characterized by low rates and aging populations. While some nations are implementing supportive policies to encourage family growth, economic factors, cultural norms, and personal choices continue to shape fertility trends. Addressing these challenges requires multifaceted approaches to support families and encourage higher birth rates.

Developing countries fertility situation Thailand's fertility rate has fallen to an average of just 1.6 children per woman, from seven in the 1970s, disrupting centuries of tradition in which children care for their parents. Other countries of the developing world have also seen sharp declines in fertility rates, including Brazil, Mexico, and parts of India and Southeast Asia. Rising prosperity appears to be one reason behind this decline. If the trend continues, the United Nations estimates that the global population will cross 8.3 billion in 2050. It will decline to 7.2 billion by 2100. A birth rate is declining in Thailand, and as a result, fewer young people are taking care of the country's elderly. Some developing nations are taking advantage of the demographic dividend. Young people will be the assets for countries economic growth. China saw its working-age population decline by 3.45 million in 2012 and 2.45 million last year, a cumulative decline of 0.63 percent since 2011. China got the success of a declining birth rate after adopting a one-child policy using carrot and stick measures, a sign that population expansion has ended. China is now relaxing its one-child policy and making it easier for people to move for the number of children they desire. South Korea, whose economy is more developed than many other countries in Asia, is trying to reduce the expenses associated with raising children. Chile's government announced plans for a "baby bonus" for parents who have a third child in 2023, following the lead of countries such as France and Australia, which already have incentives for parents to have more children.

Current Trends Declining fertility rates: According to the World Bank, the global fertility rate has fallen from about 5.0 children per woman in 1970 to around 2.4 in 2021. In many developing countries, the average fertility rate is now below 3.0, though there are significant exceptions.

Regional Variations: Sub-Saharan Africa: This region still has some of the highest fertility rates in the world. For instance, Niger has a fertility rate of approximately 6.8 children per woman (2021 data). Southeast Asia: Countries like Indonesia and Bangladesh have seen declines, with fertility rates around 2.3 and 2.1, respectively.

Specific Country Data: Nigeria: Fertility rate is about 5.3 children per woman (2021). Ethiopia: Fertility rate has declined to approximately 4.1 children per woman (2021). India: The fertility rate is about 2.0 children per woman (2021), reflecting significant progress in family planning.

Factors Influencing Fertility Rates Education: Increased education, especially for women, is correlated with lower fertility rates. UNESCO reports that for every additional year of schooling for girls, fertility can decrease by 10-20%. Access to Family Planning: A significant barrier in many developing nations is the limited access to contraceptives. According to UNFPA, about 214 million women in developing countries want to avoid pregnancy but lack access to modern contraceptive methods.

Cultural Norms: In many societies, larger families are still valued. However, attitudes are shifting, particularly among younger generations, leading to smaller family sizes. Economic Development: As countries develop economically their fertility rates decline rapidly. As per world bank indicators there is a clear link between GDP per capita and fertility rates.

Health and Mortality Factors Maternal and Infant Mortality: High rates of maternal and infant mortality can drive families to have more children. For instance, the maternal mortality ratio in Sub-Saharan Africa remains high, at approximately 533 deaths per 100,000 live births (WHO). COVID-19 Impact: The pandemic has disrupted healthcare services, including family planning. Initial data suggested a temporary decline in birth rates in many developing countries, although the long-term impacts are still being assessed. Conclusion The fertility landscape in developing countries is characterized by declining rates overall, but with significant regional and country-specific variations. Efforts to improve education, access to healthcare, and economic opportunities for women are crucial in continuing to address these challenges.

Fertility Situation in India

India's fertility situation has seen a significant decline over the past few decades. The total fertility rate (TFR) is currently around 2.2 children per woman, moving closer to the replacement level of 2.1. However, there are notable regional disparities: some states, particularly in the north, still exhibit higher fertility rates, while urban areas tend to have lower rates due to factors like better access to education and family planning services. **(See Map 4)**

CHAPTER-4

Mortality

Mortality is one of the three components of population change, the other two being fertility and migration. Historically, the factor of mortality has played a dominant role in determining the growth of population. The size of which fluctuated in the past mainly in response to variations in mortality. The increase in the population of European countries following the Industrial Revolution in the seventeenth century was mainly due to a decline in the death rates. Mortality rate has declined in the developing countries too. In fact, the single most important contribution of demography has been the revelation of the fact that sharp declines in mortality rates, rather than any rise in fertility rates, have been responsible for bringing about a rapid growth of population. The study of mortality is useful for analyzing current demographic conditions as well as for determining the prospects of potential changes in mortality conditions of the future. The public health administration depends heavily on the study of mortality, for statistics on death in the population cross-classified by age, sex, and the cause of death are of great value for the formulation, implementation, and evaluation of public health programs. It is, of course, possible to study mortality from several angles, for various biological, social, economic, and cultural factors affect the health of an individual and consequently the mortality rate in society. These factors affecting mortality can be classified under heredity, constitution, and environment. When mortality is viewed from the demographic point of view, it is studied to determine changes in the population size and structure rather than from the medical angle. The demographic study of mortality, therefore, does not usually take into consideration the genetic factor. The remaining two factors, namely, the constitutional and the environmental factors, on the other hand, provide the basis of a demographic analysis of mortality. Of the various constitutional factors, including the physical, physiological, anatomical, and psychological characteristics of man, the most important for the demographic study of mortality include the natural physical surroundings of the individual as well as his or her social and economic environments and personal habits. Reasons for past higher mortality in the past till early 19th century, the worldwide death rate was high; as a result, life expectancy was low, and population growth was less. The reasons behind it are following. Food shortage: till the Green Revolution period, food was not available for people. Due to a lack of food, people used to die. Epidemics: medical advances were less, and epidemics like plague, influenza, and cholera used to take a high toll of death of people. Recurrent wars: United Nations form to make peace in the world. A small region used to grab property and manpower of other kingdoms. In wars, people from both groups die. Sanitation: unawareness about cleanliness, use of soap, and detergent was not common; as a result, disease increased and the death rate was also increased. Natural calamities: floods, earthquakes, and volcanoes are the calamities to take a higher toll of deaths. Concepts in mortality the study of mortality deals with the effects of death on population. To assess the importance of the death in a group of people, we use the death rate.

Death: The United Nations and the World Health Organization have defined death as follows: "Death is the permanent disappearances of all evidence of life at any time after the live birth (postnatal cessation of vital functions without capacity of resuscitation)." A death can thus occur only after a live birth, and the span between birth and death is life.

Live birth is defined by the United Nations as "the complete expulsion or extraction from its mother of a product of conception, irrespective of the duration of pregnancy, which, after such separation, breathes or shows any other evidence of life, such as beating of the heart, pulsation of the umbilical cord, or definite movement of voluntary muscles, whether or not the umbilical cord has been cut or the placenta is attached; each product of such a birth is considered live-born."

Fetal deaths: any death prior to a live birth is not considered a death. Thus, abortions and stillbirths are referred to not as deaths but as fetal deaths. Any expulsion of the fetus, either spontaneous or induced, that occurs before the fetus becomes viable and is capable of independent existence outside its mother is known as an abortion.

Still birth When a birth does not have any of the characteristics included in either of these two definitions of live birth or abortion, it is known as a still birth.

Sources of Data

Vital Registration System: information about death obtained through the system of the registration of vital events. In case of developing countries, however, this system is usually very inadequate because all the events of deaths are not registered. As a result, registration data cannot be used meaningfully for an analysis of mortality. National census: In the absence of adequate registration data, the national census and demographic sample surveys provide alternate sources of information on mortality. Death rates are estimated on the basis of the available data on the age and sex distribution of the population as contained in two consecutive census operations.

Sample surveys: In sample surveys, it is possible to collect data on deaths directly by asking questions on the number of deaths in a household during a specific reference period. The Demographic Year Book of the United Nations provides statistics on the number of deaths, on death rates, death by age and sex, infant deaths, infant mortality rates, cause of death, etc., for various countries of the world. A special issue of the Demographic Year Book may give data on deaths in greater detail. Statistical reports of the World Health Organization also provide information on mortality. Recently conducted National Family Health Surveys in 1992-93 and 1998-99 provide some information on mortality in India and states. Measures of morality. The analysis of mortality uses a variety of measures.

Crude death rate (CDR) The crude death rate is the simplest and most commonly used measure of mortality, which can be quickly calculated and, at the same time, easily understood. It is a ratio of the total registered deaths of a specified year to the total mid-year population, multiplied by 1,000. It is computed as follows: It is defined as the number of deaths in a year in a geographical area per 1000 average population of the area.

It is expressed as CDR= D/P*1000 Where D represents the total number of deaths in an area during a calendar year. P represents the average population of the area in the same calendar year. **(See Map 5)**

Age-specific death rates (ASDR): Age is a very important and fundamental variable in mortality investigation. Age-specific death rates given detailed information regarding the mortality condition of a country for a comparison of mortality situation at different parts of human life. The ASDRs are the best out of all the indices.

It is expressed as nMx = nDx/nPx *K where nMX represents the age-specific death rate between ages. nDx is the death in a geographical area among a population between ages x and x+n during a calendar year. nPx is the mid-year population in the given area between ages x and x+n.

Infant mortality rate (IMR): "Infants are defined in demography as an exact age group, namely, age 'zero' or those children in the first year of life who have not yet reached age one." The infant mortality rate is generally computed as a ratio of infant deaths (deaths of children under one year of age), it is registered in a calendar year to the total number of live births registered in the same year. This rate is computed as follows:

IMR = Do / B *K. Do is the number of deaths below age one registered during the calendar year. **(See Map 6)**

B represents the number of live births registered during the same year. K is 1000. In 2012, infant mortality in India slightly declined to 42 deaths per 1,000 live births, down from 44 in 2011. Madhya Pradesh had the highest rate at 56, while Kerala had the lowest at 12. Delhi recorded 30 deaths per 1,000 live births, making it the worst among metropolitan cities.

Neonatal mortality rate (NMR): Thus, neonatal mortality is the number of deaths of infants under 4 weeks of age (28 days) or approximately one month during a year per 1000 births during that year.

NMR = NND/B* 1000 Where NND is the infant death in neonatal period.

Postnatal mortality rate (PNMR): Infants who die between 28 days and 365 days of their lives.

PNMR = PNND/B* 1000 Where PNND is the number of infant deaths in the postnatal period. D infant death [NMR + PNMR = IMR]

Maternal mortality rate (MMR) The maternal mortality rate is defined as the number of female deaths due to paternal causes among the residents of a community during a specified year per 10,000 live births.

Symbolically- Maternal Mortality Rate (MMR) = Dp/B* 10000 Where Dp is the number of deaths due to puerperal causes B represents the number of live births during the same year. India's maternal mortality rate (MMR) declined by 16% from 212 deaths per 100,000 live births in 2007-09 to 178 in 2010-12. However, this still falls short of the 2015 target of 103 deaths per 100,000 live births set by the Millennium Development Goals (MDGs). **(See Map 7)**

The crude death rate is the most widely available index of the level of mortality. A rough classification of various countries according to mortality levels may be undertaken on this basis. This index also gives us a general idea about the trend in mortality in a particular area over a period of time. Though this summary measure is a very useful indicator of the level of mortality in any population, it is not a refined measure, as is evident from its very name, and suffers from several limitations. One of these limitations arises because of the inadequate coverage of death statistics, though this is not inherent in its very nature. The crude death rates of several countries suffer from this weakness, and it is for this reason that it has to be mentioned. Another limitation of this measure is that it hides the experience of population groups whose mortality varies a great deal. On the whole, it gives a greater weight to the mortality experience of large groups in the population. As the crude death rate does not take account of the differences in the age-sex structures of different populations and as the impact of mortality is not uniform for all ages, the crude death rates of different populations are not strictly comparable. For instance, the crude death rate for England and Wales in 1961 was 11.8, while it was 11.3 for Mexico in 1960. Does this mean that the impact of mortality is the same in both countries? Some facts about the age structure of these two countries need to be known before any comparisons can be made. In 1961, England and Wales had a larger proportion of adults, that is, 77.1 percent of the population was of age 15 and above. The corresponding percentage for Mexico was 55.6 in 1960. As the mortality at the adult stage is higher, the crude death rate of England and Wales naturally tends to be higher than that of Mexico. When, however, the crude death rate of England and Wales was standardized with respect to the age distribution in Mexico, it was found to be 4.8 per 1,000 populations, a much lower death rate than that of 11.3 in Mexico. (The standardized death rate for England and Wales was computed by applying the age distribution of Mexico to the death rates of England and Wales.). To overcome the inherent limitations of the crude death rate, age-specific death rates are used. As sex differentials in mortality are also an important dimension, it is customary to compute age-sex-specific death rates. In fact, each measure of mortality is usually computed separately for comparing the mortality experiences of various populations.

The Expectation of Life at Birth

Average Expectation of Life at Birth or Average Life Expectancy: The average expectation of life at birth is a good measure of the level of mortality because it is not affected by the age structure of the population. This measure is derived from the life table, which is constructed to summarize the mortality experience of a single hypothetical generation or a cohort of people subject to a set of constant age-specific death rates throughout its lifetime. The term "average expectation of life" represents the average number of years of life to which a cohort of newly born babies (that is, those born in the same year) may be expected to live if they are subjected to the risks of death at each age according to the age-specific mortality rates prevailing in the country at the time of which the measure refers. Though the average expectation of life at birth is a measure that is rather complicated to calculate, it is most easily understood by the common man. It is, therefore, widely used to compare mortality levels and analyze mortality trends in different countries. **(See Map 8)**

Mortality rates are highest in the first hours, days, and weeks of life, with differing causes of infant deaths at early and later stages of infancy. Hence, in a study of infant mortality, infant deaths are carefully grouped into two categories according to the age at death. The first category includes infants who die before reaching four weeks of age. The other category consists of those infants who die between 28 days and 365 days of their lives. The rate based on the first period is known as the neonatal mortality rate, while that on the second period is referred to as the post-neonatal mortality rate. Factors that affect fetal and neonatal deaths are primarily endogenous, while those that affect postnatal deaths are primarily exogenous.

Endogenous Factors

The endogenous factors are related to the formation of the fetus in the womb and are, therefore, mainly biological in nature. Among the biological factors affecting fetal and neonatal infant mortality rates, the important ones are the age of the mother, the birth order, the period of spacing between births, prematurity, weight at birth, and the fact of multiple births. Of all the factors listed above, the following have been studied in great depth: the age of the mother, the parity of the mother, and the order of pregnancy or birth. It has been generally observed that fetal and neonatal mortality rates are higher at the younger ages of the mother (that is, below the age of 19), at first parity, and for the first birth order. These mortality rates start declining up to the age of 29 of the mothers and at the second and third parity and then again increase with the high birth orders of children higher age of the mother, and higher parities. Thus, if a graph of fetal and neonatal mortality rates is drawn with respect to these factors, it would more or less resemble a U-shaped curve. The maturity of an infant at birth has also been found to be an important factor affecting neonatal and infant mortality rates. It may be concluded from this discussion that the causes of fetal and neonatal deaths so far considered arise mainly out of genetic factors and may be traced back to the intrauterine life of the fetus and to the damage occurring during the process of birth.

Exogenous Causes: Social, cultural, economic, and environmental factors are also found to affect infant mortality, especially during the post-neonatal period. Post-neonatal deaths are therefore mainly due to various epidemics caused by communicable diseases, both of the digestive system, such as bronchitis and pneumonia, as well as by faulty feeding patterns and poor hygiene. The underlying environmental factors include crowding and congestion, insanitary surroundings, lack of proper sunshine and fresh air, etc. Suppositious is also an important factor contributing to a high infant mortality rate. The difference between infant mortality rates of legitimate and illegitimate births is usually found to be quite marked. the reason for the difference is obvious. A child conceived and born out of wedlock is generally unwanted both by the mother as well as society. Consequently, such a child does not receive Factors influencing mortality.

Marital Status and Mortality: Mortality is low among married people. This is because the possibilities of getting married are high among those who are physically fit. Besides, mortality among married persons is low due to the satisfaction in married life and the protected life a married person lives. As against this proportion of physically fit persons among unmarried is relatively low. Besides, unmarried persons live a carefree life. Mortality among unmarried persons, therefore, is relatively high. Mortality among unmarried persons, therefore, is relatively high. Mortality among married women is low due to the satisfaction gained due to the childbirth.

Nature of Occupation and Mortality: Mortality is closely linked to the nature of the occupation and the nature of the place of work. Similarly, mortality is also related to nature of occupation, socio-economic status, and standard of living. Different risks are involved in different occupations. The proportion of those dying due to heart diseases is high among white-collar workers. Similarly, the proportion of those dying due to tuberculosis is high among those working in mines and textile units. The proportion of those dying due to accidents is high among worker's transport.

Urbanization and Mortality: Initially, it was difficult to keep cities due to a lack technical development. The death rate, therefore, was high among city dwellers. For example, in 1841, the expectancy of life for males in London was 35 years. Whereas the expectancy of life for males in the rest of England was 40 years.

With improvements in water supply, drainage disposal, and medical facilities, cities became cleaner. The death rate, therefore, declined, and expectancy of life among city dwellers increased. For example, in 1952, the expectancy of life for males in London was 67 years, while the expectancy of life for males in the rest of England was 66 years. In India, in 1967, mortality for rural areas was 13 per thousand, while for urban areas, mortality was 8 per thousand.

Education and Mortality: Education has an inverse relationship with mortality. Educated people have better social and economic status. That is why mortality among them is generally lower. For example, a study in Bombay city showed that there is a close relationship between the education of a mother and the infant mortality rate. Infant mortality rate was highest among illiterate mothers. As against this, the infant mortality rate was very low among children born to mothers who were educated up to matriculation and above.

Mortality in India

In any study related to mortality in India, the absence of reliable data is a major hindrance. It is very difficult to get mortality precisely because of a lack of proper recording. Even then, from whatever records are available, we can say that major changes occurred in mortality during 1871 and 1921. These changes occurred due to famines, epidemic diseases, and wars. Severe famines were experienced in India from 1876 to 1878 and again from 1898 to 1900. In India, the influenza epidemic was rampant during 1918. 15 million people died due to the influenza epidemic. Because of this death rate, which in 1917 was 33 per thousand, it increased to 63 per thousand in 1918. This epidemic was brought under control after 1918. Mortality, therefore, started declining. Mortality also declined due to improvements in agricultural production and improvements in the land-distribution system.

The crude death rate at all India levels has declined significantly from 14.9 to 12.5 during 1971 to 1981 and thereafter from 9.8 to 7.2 during 1991 to 2011. The decline is less in rural areas than in urban areas. The infant mortality rate, is plays a vital role in health planning, has shown a considerable decline from 129 per 1000 live births in 1971 to 110 in 1981 and from 80 in 1991 to 47 in 2011. The child mortality rate has observed a perceptible decline from 51.9 in 1971 to 41.2 in 1981 and from 26.5 in 1991 to 13.3 in 2011. In 2011, about 34 percent of the deaths were institutional, and 66 percent received medical attention other than institutional. The crude death rate (Statement 41) at the national level for 2011 is 7.2 per thousand population, and it varies from 7.7 in rural areas to 5.8 in urban areas. Among the bigger states, it varies from 4.2 in Delhi to 8.6 in Odisha. The states having a death rate higher than or equal to the national level are Andhra Pradesh (7.6), Assam (8.2), Chhattisgarh (8.0), Madhya Pradesh (8.3), Odisha (8.6), Uttar Pradesh (8.1), and Tamil Nadu (7.6).

Infant Mortality The percentage of infant deaths to total deaths varies from a low of 2.8 percent in Kerala to a high of 21.8 percent in Rajasthan. Percentage age contribution of deaths from people's 60 years and above varies from 37.5 in Assam to 71.2 in Kerala. The percentage share of infant deaths to total deaths by residence in 2011 was 14.5 percent and varies from 15.8 percent in rural areas to 9.7 percent in urban areas. In rural areas, Kerala registered 3.0, the lowest percent share of infant deaths as compared to 24.5 in Rajasthan. In urban areas, the difference is from 2.3 percent in Kerala to 17.1 percent in Uttar Pradesh. Infant Mortality Rate (IMR) by sex and residence for the year 2011 for India and bigger States shows that at the national level, IMR is reported to be 47 and the variation is from 51 in rural areas to 31 in urban areas of India. Among the bigger states, it varies from 13 in Kerala to 62 in Madhya Pradesh. Infant mortality among Female is higher than that of male infants in all states of India.

CHAPTER-5

Migration

Introduction The movement of a person or group of people is called migration. The main question about migration is, why do people migrate? To answer this question, different scholars in this field coined models, namely Zipf's gravity model (1946), Stouffer's model of intervening opportunity (1946), Brown and Moore's model of activity space and urban residential mobility (1972), Golont's model of migration as an adjustment process (1972), and many others. Most of these, however, refer to urban to urban residential mobility, especially in the developed countries, and, as such, are not discussed here. Furthermore, there are also numerous deterministic and probabilistic models of migration, with varying degrees of mathematical or statistical sophistication, that primarily deal with a completely different theoretical issue: how far people migrate? Evidently, such models lie beyond the scope of the present discussion. Our present concern is to understand the reasons for: Why people move?

Migrant: A Migrant is a person who changed place of residence from one migration defining area to another at least once during the migration interval (interval is one year, five years, or ten years, or intercensal period).

Place of origin: the place or area from which a move is made is the place of origin. The place of origin is an area of residence at the beginning of the migration interval, or an area of residence from which last move was made for the current migration interval.

Place of Destination: The area in which a movement terminates is the place of destination. Migration Stream: the total number of moves made during a given migration interval that have a common area of origin and of destination.

Lifetime Migration: the place of residence at the census/survey date is different from the place of birth; this is called a lifetime migration.

In-migration: Every move is an out-migration with respect to the place of origin and an in-migration with respect to the place of destination.

In-migrant: A person who moves into a migration-defining area from within the same country. He is to be distinguished from an "immigrant" who is an "international migrant" entering the country from a place outside the country.

Out-migration: An out-migrant is a person who reports from a migration-defining area by crossing its boundary to a point outside it, but within the same country. He is to be distinguished from "emigrant," who is an "international migrant" departing to another country by crossing an international boundary.

Gross and Net Migration: Data that refer to all moves or all migrants, within a specific definition of migration that is being applied, are concerned with "gross" migration. With respect to a given area, the sum of in-migration and out-migration is referred to as gross migration, and of in-migrants as gross migrants.

Net migration refers to the balance of movements from both the directions. With reference to a given area, it is the difference between the volume of in migration and outmigration. When in-migration exceeds out-migration, the net going to are is called "net in migration," which takes a positive sign. In the opposite case, there has been "net outmigration," which takes a negative sign.

Measures of Migration

Internal migration: Internal migration for measuring or estimating in two categories: Direct techniques based on particular question. Indirect techniques, it does not require special questions. The direct techniques are based on the movement of the person. Direct measures of migration.

Place of births. Duration of residence Place of last residence the location where an individual lived at a specific date in the past, such as one or five years before a census. Indirect measures of migration.

The vital statistics method: The survival ratio method Migration rate Migration Rate: Example of indirect measures: Is the number of migrants related to the population that may have migrated during a given migration interval?

The migration rate is computed as follows:

M = M/P *K Where M is the rate of migration for the specified migration interval. M is the number of migrants or the number of people's migrations during the interval. P is the population exposed to the likelihood of migration during the interval. K is 100 or 1000.

Types of Migration

There are two major types of migration: internal migration that occurs within a country and international migration that takes place across international boundaries. The processes, causes, and consequences of internal migration are much different from those of international migration, as the former refer to the socio-economic spatial situations within a country, while the latter refer to the international socio-economic and political conditions, especially immigration and emigration laws and policies of these countries.

Rural-Urban Migration: Within internal migration, there are four-way classifications of migration according to their direction of movements within and between rural and urban areas. These are Rural to rural migration Rural to urban migration Urban to rural migration Urban to urban migration of all these streams, it is primarily rural to urban migration that becomes the most important, as it contributes to the transfer of labor force from the traditional agricultural sector to the urbanized industrial sector and is liked with the process of urbanization. Likewise, urban migration is also related to the process of concentration of population in large towns and cities. However, in many developing countries like India, rural to rural migration is also of great importance and significance, especially among the females, who move primarily due to marriage or for familial reasons. Each of these migration streams has their own reasons and repercussion.

Push-Pull Hypothesis

According to this hypothesis, there are two factors that are to be considered together for understanding the causes of migration. These are the push factors at the origin places and full factors at the destination places. R. Herberle (1938) and J. C. Mitchell (1946) have elaborated on this conventional concept of push-pull factors. These factors have been used to explain internal migratory movements of the rural-rural, rural-urban, or urban-urban types as well as international migrations. Push factors at origin and pull factors at destination: Factors such as unemployment, underemployment, lack of adequate cultivable land, or lack of job opportunities and basic amenities of life (like school, college, or hospital), or socio-economic-political stresses or natural hazards (like floods and draughts pull factors at the destination place., draught, etc.) out migrate places often tend to push people to out-migrate from their places. On the other hand, better job opportunities, better prospects, higher wages, school/college facilities, better amenities, health services, etc. attract people towards a particular destination. These are the pull factors at the destination place. Evaluation: In spite of the strong reaction for such simplistic formulation and explanation of migration, the push-pull hypothesis has gained considerable scope in almost every migration literature. It is said that migration is the result of the balancing of these two opposite forces.

Ravenstein's Laws of Migration: In the late 19th century, E.G. Ravenstein postulated a number of laws on migration based on data collected from more than 20 countries, including the British Census of 1881. His two papers of 1885 and 1889 formed the starting point for both empirical and theoretical work on migration that continues to be relevant even today. Ravenstein's laws of migration, in brief, are as follows: Migration and distance: Majority of people migrate for short distance. If migrants proceeding long distance generally go by preference to one of the great centers of commerce and industry. Migration by stages: Migration occurs in the following stages: the inhabitants of the rural area, immediately surrounding a town of rapid growth, move into it, and the gaps thus left are filled up by migrants from still remote areas, until the attractive force of one of the rapidly growing cities makes its influence felt. Stream and counter-stream: Each main current of migration produces a compensating counter-current. Urban-rural difference in propensity to migrate: The natives of towns are less migratory than those of the rural parts of the country. women are more likely to migrate within a country than men, especially for short distances. Technology and migration: An increase in the means of locomotion and development of manufacture and commerce leads to an increase in migration. Main reason for migration is economic. Evaluation: Ravenstein's laws formed a pioneering work and have survived over a period of hundred years, but these are simplistic laws that neither portrayed the entire truth nor did they penetrate into the underlying socio-economic and political forces that were (and still are) the real reasons for such massive migration. The main shortcomings of these laws are: They do not explain why great centers of commerce and industry are formed, neglecting the growth of other centers, and why migrants are forced to migrate to them while passing others. In developing countries, it is seen that migrants make a quantum jump to the larger metropolises (e.g., in the case of India, to metropolises like Delhi, Mumbai, Kolkata, Chennai, etc.) by passing smaller towns. So, the law of stage-wise migration might have described the situation in the advanced countries. As regard stream and counter stream, it is not always true that a stream from I to j is followed by a counter stream of migrants from j to I. For example, migrants from a village to Bombay city are not always being followed by a counter-stream of migrants from Bombay city to that village. It is not because of the 'propensity' (natural tendency) to migrate that causes massive rural to urban migration, but rural poverty, which in reality forces rural inhabitants to migrate, while urban residents remain largely immobile or move for a better future in their lives. The predominance of females among the short-distance migrants is true where marriage migration is considered, but it does not hold true in the case of labor migration.

The relationship between technology and migration appears to be true, but there is no reference to the effects of markedly uneven spatial characteristics of the development of industries and transport upon migration. This is because of the result of colonization and lopsided/polarized investment made only at a few chosen centers in a country, neglecting the development of other centers or regions. The economic motive is the most important factor in migration, but the way this law is stated, it does not leave any scope to explain those huge streams of migrants that arise in the developing countries out of poverty, hunger, underdevelopment, and rural stagnation.

Lee's Theory of Migration (1965): Evertt S. Lee, at the annual meeting of the Mississippi Valley Historical Association in 1965, presented his theory of migration. According to Lee, the factors that affect the decision to migrate and the process of migration are: Factors associated with the area of origin, viz., higher wages, more job opportunities, etc. Factors associated with the area of destination, viz., risks, uncertainty, and expectation at destination; Intervening obstacles between the area of origin and destination, viz., distance, cost, restrictive laws, etc.; and Personal factors, viz., age, sex, race, education, skill, etc. These factors can have plus, zero, and minus values as far as the role that they play in the movement is concerned. It is not only these factors but the perception of these that are, according to Lee, important in the decision to migrate. Based on these factors. Lee postulated sets of laws regarding the volume of migration streams and characteristics of migration selection.

CHAPTER-6

Marital Status (Nuptiality)

Introduction

The term "nuptiality" is derived from the word "nuptial," which means marriage or wedding. Demography, "Nupatiality" relates to the study of nuptial unions as well as of nuptial dissolutions. The formation of nuptial unions may take place either by first-time marriages or through remarriages, and the dissolution of nuptial unions may occur by divorce or widowhood. Marriage and widowhood are the important demographic events that influence the course of population growth. Nuptiality is a topic of interest to demographers in its own right, but largely it has been singled out for intensive study because of its impact on fertility. The changes in population size due to births could be considered as a function of one generation producing another, which takes place only within marriages for most societies. Among the various variables determining the formation and dissolution of nuptial unions, the female age at marriage and the proportion of those who are never married in the reproductive age group are considered to be important. The female age at marriage and its effect on societal fertility is an investigated that fertility is strongly linked to the length of exposure in nuptial unions. Topic in the field of demography. It is believed that fertility is strongly linked to the length of exposure in nuptial unions. The proportion of the population dissolved, the age at affect it marries, and the extent to which marriages are dissolved; it can affect the birth rate or level of fertility. At the present time, the birth-rate is the primary determinant of population growth in most developing countries. We can study the trends and differentials in nuptiality just by looking at the proportion of persons in different categories of marital status in each age group and their patterns in space over time. The pattern of nuptiality varies sharply with age; children and adolescents tend to be single, young adults and adults tend to be married, divorce tends to be the greatest at the adult ages, and widowhood is concentrated at older ages. Although these generalizations tend to apply to all populations, there are variations between the populations. In almost all the societies of the world, childbearing is allowed after marriage. Thus, the proportion of married persons in a population has profound effects on fertility behavior. Fecundity, i.e., the biological capacity of a woman to reproduce children, is not uniform across the entire length of her reproductive span. Reproductive span in the life of a woman commences from menarche, i.e., the first menstruation in the early or mid-teens, and terminates at menopause, i.e., the permanent cessation of menstruation that occurs, on average, in her late 40s. Evidence indicates that the fecundity level of a woman, which is at its lowest level in the early years of her reproductive span, undergoes a 6-7 fold increase in her 20s and again tends to decline, only to stabilize at a very low level in her late 40s. An early marriage of woman, thus, means not only a longer duration of childbearing but also a more frequent childbirth. Age at first marriage of a woman, became crucial phenomena for fertility level in a population. In societies where marriage is a universal phenomenon and where the contraceptive prevalence rate is very low, rising age at marriage holds a prime position in population control programs. Similarly, marital status exerts a significant influence on the health and morbidity conditions of an individual. Based on data from some European countries, Cox has shown that bachelor's mortality exceeded that of married men by a considerable margin in all ages. The end of marriage by the death of one of the partners brings about a sudden change in the mode of living of the survivor, whose vitality is temporarily influenced and may even be (Cox, 1975:121). Thus, both brave men and women suffer poorer health than those still married, and the mortality rate among them is invariably higher. Also, marriage of woman, particularly when it occurs in the initial years of her childbearing span, exposes her to extra risk to health from childbearing, sometimes leading to death. The less developed societies, characterized by the practice of early marriage, report a significantly higher incidence of maternal mortality. Further, infants born to very young mothers have a much higher probability of dying during the first few weeks or months. The National Family Health Survey (NFHS), in its second round, reports that neonatal mortality (within the first four weeks of the birth) of infants born to mothers below 20 years of age in India is higher by nearly 42 percent than among those born to mothers aged 20-29 (IIPS, 2000:192). Even child mortality rate (up to 5 years of age) is higher in cases where mother's age at childbirth is less than 20 years. Thus, age at marriage in a society plays a very crucial role in the mortality condition of children and infants. Finally, changes in marital status of individuals induce migration. Although not much is known about the precise link between marital status and migration, marital status effect on the spatial mobility of the population. In patrilocal societies, brides move to grooms' places after marriage. Such societies, therefore, exhibit greater mobility among females than males. In India, females are more mobile than males. Migration is, at the same time, highly age-selective. People in the working age groups have a higher propensity to migrate. A bachelor is more likely to migrate in search of job opportunities or better education facilities than a person who is tied in a marital bond. In some cases, for instance in India, the end of marriage by death of the spouse, particularly when it occurs soon after marriage, often results in the return of the widow to her parent's place.

Sources and Quality of Data Sources of data:

There are three important sources of information in nuptiality statistics. These are: Vital registration system National census Sample surveys.

Vital Registration System: Since marriage is a vital event, data on marriage are collected in vital statistics. But the extent of completeness varies from country to country. Few countries have a good registration system, but a majority have poor data quality from this system. In most of the developing countries, registration of marriage is highly deficient, and thus the data on marriage statistics collected through this source are meaningless. The Demographic Year Book of the United Nations, published periodically, contains information on registered numbers of marriages, divorces, and remarriages for various countries. National Census: The data on "marital status" are usually collected through censuses for most countries. The Census report generally publishes this information by age, sex, and marital status, which is of great use to the demographers. Moreover, it is a regular feature and thus enables us to examine the time trend. While using the census returns over a period of time, one has to examine a few basic items before analyzing. Boundaries of a country or region may not be the same over a period of time. The boundaries are the same, but the area covered by the census is not the same over a period of time. Apart from the adjustments due to boundary changes, examination of the marital status distribution is also necessary.

Sample Surveys: Demographic sample surveys, especially fertility surveys, collect information on marital status, age at marriage, age at consumption of marriage, dissolution of marriage, remarriage, etc. The data from surveys usually provides fairly reliable information and has several advantages over the vital registration or the censuses. The main advantage is to collect a great deal of information for analysis of nuptiality. Although sample surveys are not a regular feature, for several countries they have been conducted wherein information on marriage statistics is available. At the international level, sample surveys like the World Fertility Surveys (WFS) and Demographic Health Survey (DHS) have collected a large volume of demographic data on nuptiality. In India, many surveys conducted in the past contain information on marriage statistics. The most important among them are the baseline surveys conducted in the various states during the decade 1980-90. The sample registration system (SRS) has also collected data on nuptiality, which is available from various publications. Apart from these, there are parish records, which also provide data on marriage statistics. Church registers at many places have data that is important for historical studies of nuptiality.

Quality of Data: The data on marital status follow some trend with age, and any deviation from the general trend puts in suspicion. For example, the proportion of single persons decreases with age; the proportion of currently married persons first increases with age, and then it decreases slowly. Similarly, the proportion of widowed persons increases with age. If there is any deviation from this general trend, it is worthwhile to examine the cause of the deviation. If necessary, adjustments should be made for misstatements of age and marital status.

Measures of Nuptiality

Information on the marital status of the population in a country is obtained in its periodic censuses. The data is usually obtained under five broad categories recommended by the United Nations. These categories are as under: Single persons, i.e., never married, currently married persons, Persons divorced and not remarried, Persons widowed and not remarried, and Persons legally separated.

Concepts and Definition of Terms

Marriage: refers to the legal union of persons of the opposite sex. The legality of a union can be established through civil, religious, or other methods as recognized by each country's laws. The phenomenon of marriage may be classified into two types: First marriage refers to persons moving from singlehood (bachelor or spinster) to married status, and Remarriages refer to persons moving from divorced or widowed status to married status.

Widowhood: Refers to the dissolution of nuptial union upon the death of one of the spouses. The male partner upon the death of wife is termed as widowed, and females as widowed upon the death of husband.

Divorce refers to the final legal dissolution of a marriage, which means separation of husband and wife by a judicial decree. The legal marriage may be broken in three ways: either by (a) divorce decree, (b) death of spouse, or (c) annulment.

Annulment: Refers to the sanction provided in very few societies/countries in which the marriage contract becomes void from its inception, whereas a divorce makes the contract void as of the date of the decree. Annulment confers upon the parties the status of never having been married or annuls of marriage that have taken place earlier.

Consensual Union: Refers to a marital union without recorded legal sanction. In some Central and South American countries, this type of union constitutes a large proportion of all marriages.

Marital Status: The population of any country can be classified into different status of marital unions or dissolutions.

Almost all cultures of the world recognize four categories of population in different marital status. These are, never married or single persons, currently married persons, Widowed/widowered persons, and divorced and separated persons. Each marital status has a deep religious and legal as well as general social significance. It is necessary to seek legal permission (or societal sanction) for being married or divorced in most societies. Also, most religions require religious ceremonies or declarations for change in marital status. The marital composition of the population may broadly be categorized into two parts: Never-married persons, and Ever-married persons. The first group, "never-married persons," refers to only single persons and consists of all persons who have not married (never married) earlier, i.e., bachelors + spinsters. The second group, "ever-married" persons, refers to tall those who have been married for at least one time, although presently they may/may not be married. This group consists of currently married persons, widowed persons, divorced persons, and separated persons. The above classification of marital status is important in the study of nuptiality and helps in describing and understanding the pattern and characteristics of population. **(See Map 9)**

The United Nations (U.N.) has recommended the following five categories for marital status:

- Single (never married)
- Married and not legally separated
- Widowed and not remarried Divorced and not remarried
- Married but legally separated.

The above classification of marital status is to make each status very specific in concept and in meaning. In most of the countries, the last two categories are combined since the proportion of population in this category is usually not very significant. In some countries, especially in Latin America, a unique status of married category is "in consensual union," or living together. Systems of classification by marital status vary from country to country in accordance with the prevailing law and customs. Marital status is an important aspect of the population composition of any country. It affects labor force participation, school attendance, rural-urban residence, and almost every other topic in which demographers are interested. The last of these categories put together represents 'ever married' persons in a population. In addition to these five categories, sometimes yet another category is added for the persons who do not report their marital status. For example, in India, a sixth category on 'not specified' can be seen in the tables on marital status. In India, data on marital status is provided in the 'Social and Cultural Tables.' In some countries, where extra-legal consensual unions are common, additional categories are incorporated in the census counts. Apart from the periodic census, various demographic surveys—especially fertility surveys—also provide data on marital status in specific populations. The Demographic Yearbook, the annual publication of the UN, provides data on marital status for different countries collected by periodic censuses and other sources. The data on marital status of the population is usually given for males' females separately by broad age group starting from some minimum age. In India, up to the 1951 census, data on marital status was compiled for all age groups, including children. However, from the 1961 census onwards, persons below 10 years of age are returned as single, irrespective of their actual marital status. The most commonly used measure in the analysis of marital status is the percentage distribution of population in different categories. However, from the point of view of marriage patterns, it is more useful to look into the proportion of singles in some early age groups. A change in the proportion single over time indicates the trend in the average age at marriage. The average age at marriage can be separately worked out for a population, where the measure is known as the mean age at marriage. Mean age at marriage, i.e., the average age at which individuals in a population get married, refers to cohorts of persons. The calculation of mean age at marriage, thus, requires data on the exact age at marriage of individuals in a cohort. From the periodic census, information on the exact age at marriage is not available. A widely used variant, therefore, is what is known as singulate mean age at marriage (SMAM). The measure uses census data on the proportion of persons never married in each age group to create a synthetic estimate of the mean. The measure, thus, provides a group of individuals in a population lived single for the average number of years. First proposed by Hajnal in 1953, the measure is based on certain assumptions. These assumptions are: That the population is closed to migration, and that in the population there are no mortality differentials by marital status. That the age pattern has not undergone any abrupt change over time. Agarwala (1962) later suggested that migration and mortality differentials do not cause any serious error in the estimates. The measures are, therefore, very useful for making comparisons across different groups or regions (Goya R.P., 1982: 109)

Hajanal's methods of calculating the singular mean age at marriage are very simple and can be expressed as follows:

SMAM = [nSx-Sk (K)]/ 1-sk

Where nSx is the proportion single in the age group x to x+n, K is the upper limit of the age under which marriage occurs (in most of the studies this limit is taken as 50 years), Sk is the proportion single at age k, and n is the interval of the age groups. The above-mentioned two commonly used measures can be worked out for both males and females separately. However, in most of the studies, measures on marital status are generally confined to females only, as it is more appropriate for the analysis of demographic characteristics.

Determinants of Marital Status: The distributions of population in different categories pertaining to marital status and the mean age at marriage in a population depend on the interplay between a large number of social, economic, and cultural factors. The universal marriage is common practice in India and has its roots in religious prescriptions. In the case of Hindus, who constitute over 82 percent of the population in the country, the principles of *'Purusarthas'* prescribe four stages in the life of an individua *Brahmcarya, Grihastha, Vanprastha, And Sanmnyasa*—each having its own duties and functions (Kapadia, 1966:27). For Hindus, marriage is essential for a man to accomplish the duties of *Grahasthashram*—the second stage of life. For a woman, marriage is essential because it is the only sacrament that she is allowed in her lifetimes, unlike a man who goes through several sacraments (Bhende and Kanitkar, 2000:162). Cultural practices have traditionally favored an early marriage in India, as a result of which 'proportion single' particularly among females in the age '15 and above' is one of the lowest in the world. Similarly, some religions prescribe strict sanctions against divorce. In countries inhabited predominately by Roman Catholics, for whom divorce is not permitted, one can see a very insignificant proportion of 'divorced and separated' in the population. As shall be seen later, the countries in Western Europe have been marked with unique marriage patterns, which find their roots in social and economic conditions and their historical experiences. The example of Ireland will suffice to illustrate this. The land and establishment of an independent house are preconditions for an individual in the agrarian population for marriage. Land could be acquired through inheritance or through reclamation of wastelands. Laws governing sub-divisional of land in Ireland and also in many other European countries were strict. Inheritance was permitted to only one son usually the eldest one and the rest of the sons were required to seek employment outside. This practice was known as the institution of primogeniture (Bhagat, 2002:19). The ones who were not allowed to share in their father's land were compelled to delay marriage till they got jobs. Even the eldest son had to wait till such time when the father was willing to resign both authority and property. As a result, the age at marriage for both males and females in the country became one of the highest towards the middle of the nineteenth century after the institutions were reinforced. On the contrary, in European Russia, where land was in abundance, marriage of young persons was a precondition for getting land. Landowners were interested in maximizing labor supply by encouraging their serfs to marry their children early. Scholars are of the opinion that the system of joint and extended family has been responsible for early marriage in many of the developing societies of the world. It is also argued that late marriage has been associated with nuclear and conjugal family systems. In societies characterized by a joint family system, an individual is not required to make independent arrangements of housing and life, and the responsibilities of maintaining a family are shared by all family members. As against this, in the nuclear family system, an individual is required to be in a position to provide support independently, which ultimately delays entry into wedlock. The joint family system has never been a feature of the European countries. On the other hand, in most of the developing societies of the world, the joint or extended family system has been a predominant feature. The weakening of systems during recent time has, therefore, been accompanied by a rise in the age at marriage in such societies. Among other social and economic determinants of marriage patterns, mention may be made of literacy and employment levels among females, prevalence of the dowry system, economic status of the people, etc. In a developing country like India, literacy and levels of educational attainment among females appear to be a very important determinant of age at marriage (Malaker, 1978; Pathak, 1980; Pandey, 1984). Similarly, engagement of women in gainful employment may affect their average age at marriage. In addition, in India, although there are differences of opinion, some scholars have suggested inter-caste differentials in age at marriage. Though the European experiences have been different, evidence indicates that improvement in material prosperity in the less developed parts of the world has led to a rise in age at marriage.

Among the demographic factors that influence marriage pattern, levels of urbanization, mortality rates, and numerical pattern, levels of urbanization, mortality rates, and numerical balance between male and female in the population are important. The process of urbanization erodes old practices and traditions and brings about a new set of social values that are amenable to industrial societies. Again, though the European experiences were different, in the less developed societies, the process of urbanization has definitely led to a rise in age at marriage. In Europe, even during the nineteenth century, the age at marriage was higher in rural areas than in urban areas. In the non-European realm, social values, customs, and practices in rural areas invariably favor an early marriage. Several studies have established a marked rural-urban differential in age at marriage. In India, too, the instances of early marriages are more common in the rural areas. It has also been suggested that one of the reasons for early marriage in many societies in the past has been the high mortality rate. To maintain population size, early marriages were preferred in the society. Decline in mortality rates thus motivated delay in marriage. As a result, marriage depends on availability of men as women in the marriageable age group in the population. According to Hajnal (1965) mortality and migration affect the European marriage pattern. In India, a growing imbalance in sex ratio over the past few decades in its northwestern parts has compelled people to seek brides from other regions during the recent years. The average age at marriage is also likely during the recent years. The average age at marriage is also likely to have undergone an increase in the areas marked with a deficit of females in the marriageable ages. World patterns: World patterns in marriage have historically been characterized by a marked contrast between the European realms and the rest of the world. In one of the earliest studies, Hajnal (1965) explain the world patterns in marriage into two types European and non-European. The European type has been attributed with a higher age at marriage and significant proportion of single people in the population. In contrast the non-European type, has been characterized by a universal marriage practice and a low age at marriage. The 'proportion single' in the population was normally above 10 percent, and in some countries, it was as high as 25 to 30 percent. Based on the marriage index, Coale (1973) has also highlighted the difference between the European and non-European countries. Coale was also of the onion that pattern. France and Spain significantly differed from the typical marriage trends seen in other European countries. Sklar (1974) has, however, cited examples of a reasonably higher age at marriage even in some East European countries, which were earlier classified into non-European types. The marriage pattern in Europe has undergone a marked change during recent times. After the end of the Second World War Female age at marriage in many European countries declined significantly. The marriage rate has been increased the post-war period (the marriage boom) by many scholars (Bhagat, 2002:15). The magnitude of regional variation in age at marriage significantly narrowed down as compared to what prevailed towards the close of the nineteenth century. Outside Europe, as revealed by World fertility survey data, the female mean age at marriage in Africa is around 18 years. The Latin American Counties report a higher mean age at marriage for females and also a lower range of differences from one country to another, as compared to Africa. Countries in Asia report a much wider variation in age at marriage the lowest being just 16 years in Bangladesh and the highest being over 24 years in a developed country like Japan. The same scenario happened in Sri Lanka. A higher female age at marriage observed in the Countries of the eastern parts of Asia than that in Southeast Asian countries. In Southeast Asia, female mean age at marriage varies from 19 years in Myanmar to 23 years in the Philippines and over 24 years in Singapore (Bhagat, 2002:16). India reports one of the lowest female ages at marriage in the world. So far as the relation between economic development and age at marriage is concerned, European experiences are different from that in the non-European countries. Evidence indicates that among the European countries. Evidence indicates that among the European countries, age at marriage was higher even prior to the industrial revolution. In many of the European countries, on the other hand, age at marriage recorded an increase during the time of economic depression in the 1930s and during the war periods. When the situation improved, age at marriage again declined. The same trend was experienced in other non-European countries like the United States, New Zealand, and Australia, which had close ties with Europe in terms of the origin of their populations.

India's Case

The Census is the main source of data on marital status in the country. In addition, the National Sample Survey (NSS) and various demographic sample surveys provide estimates on marital status in the country. Information on marital status for periods prior to census taking in the country is purely descriptive and is based mainly on religion text (Bhagat, 2002:59). Agarwala (1962) made one of the earliest attempts on the study of marriage patterns in the country. He derived the estimates on 'proportion single' and female mean age at marriage based on census data for the period 1891 to 1951. This was followed by similar attempts by many other demographers. One of the latest studies on the marriage patterns in India is by Bhagat (2002). The study has the distinction of being the first attempt by geographers. Bhagat has presented a very comprehensive account of early marriages and their social and economic correlates in India from a spatial perspective.

Table 1: Proportion Single and Singulate Mean Age at Marriage of Females in India, 1891-1991

Census year	Proportion Single				SMAM (years)
	5.9	10-14	15-19	20-24	
1891	0.87	0.49	0.13	0.03	12.5
1901	0.89	0.54	0.15	0.04	13.1
1911	0.89	0.55	0.16	0.04	13.1
1921	0.91	0.60	0.19	0.05	13.7
1931	0.86	0.49	0.17	0.04	12.7
1941	0.92	0.76	0.25	0.04	14.7
1951	0.95	0.83	0.28	0.04	15.6
1961	*	0.81	0.29	0.06	15.9
1971	*	0.88	0.44	0.10	17.1
1981	*	0.93	0.56	0.14	18.3
1991	*	0.95	0.64	0.17	18.9

Note:* from 1961 onwards all persons below 10 years of age have been treated as never married irrespective of their marital status.
Source: Bhagat, table 3.1, p.62.

As stated earlier, India's population has since long been marked with one of the lowest ages at marriage in the world. It is important to note that till sometime in the recent past, child marriage was a widespread practice in the country. Even in modern times, instances of child marriage are reported in the media despite the fact that marriage below the minimum age (21 years for boys and 18 years for girls) is a cognizable offense according to the amended Child Marriage Restraint Act of 1978. Scholars are of the opinion that the practice of child marriage in the country has its origin only in the seventh century. Evidence indicates that the average age at marriage in the country was quite high during the Vedic period. Kapadia suggested that the practice of child marriage in the country began sometime in the seventh century under the influence of religious writing such as *Dharma sutras* and *Smritis.* These religious texts prescribed marriage for daughters before they attained puberty. According to these writings, parents who fail to give their daughters in marriage before puberty will go to hell. The importance of virginity was also highlighted as the virtue in a woman. Thus, keeping a daughter unmarried after she attained puberty increasingly became a matter of anxiety for the parents. Under the influence of such writings, child marriage spread fast and took deeper roots in our society (Kapadia, 1966). It flourished throughout the medieval period and further degenerated during the British period, taking the shape of infant marriage. Table 1 presents the trends in the 'proportion single' women in India in some of the younger age groups, viz., 5-9, 10-14, 15-19, and 20-24 years for the period 1981 to 1991. The table also gives the estimates on singulate mean age at marriage (SMAM) for females during the period. As the table reveals, the marriage pattern in the country has undergone marked improvement during the last century. The 'proportion single', which was only 49 percent in the age group 10-14 years at the time of the 1891 census, has increased to 95 percent in 1991. Similarly, while only 13 percent of the females aged 15-19 were single in 1891, the figure was as high as 64 percent at the time of the 1991 census. In fact, the 'proportion single' has recorded a continuous increase across all age groups during the period, expecting on the occasion of the 1931 census. In addition, the 1961 census also reported a decline, albeit marginally, in the proportion of singles in the age group 10–14 years. Barring these two aberrations, the SMAM for females has reportedly increased from 12.5 years in 1891 to 18.9 years in 1991. Thus, over a period of one hundred years, the average, the gain in the 'proportion single' as well as SMAM has been more conspicuous during the post-independent period. Age at marriage for females in the country has gone up by 6.4 years. Remarkably, the gain in the 'proportion single' as well as SMAM has been more conspicuous during the post-independent period. The decline in the 'decline in the 'proportion single' in all groups and in SMAM on the eve of the 1931 census can be attributed to the sudden rise in the child marriage rates towards the close of the 1920s. It may be recalled that the British government had passed the Child Marriage Restraint Act in September 1929, which became effective only in April 1930. The intervening period between the passage and enforcement of the Act witnessed large-scale marriages involving even infants to escape from the act. Based on the state-level trends in 'proportion single' women below 14 years of age, Bhagat (2002) mentioned that at the turn of the twentieth century, the states of Andhra Pradesh (formerly Hyderabad), combined Bihar and Orissa, West Bengal, combined Maharashtra and Gujarat (formerly Bombay Presidency), Uttar Pradesh (formerly United Provinces), and Madhya Pradesh were marked with a significantly higher incidence of child marriage than the nation's average. Rajasthan joined the group in the post-1931 period. Although the 'proportion singles' is found to have increased across all the states, Andhra Pradesh, Madhya Pradesh, Rajasthan, Uttar Pradesh, and Bihar continued to be on the lower side. Rajasthan, Uttar Pradesh, and Bihar continued to be on the lower side up to 1971. The incidence of child marriage declined faster thereafter.

Table 2: Proportion Single Women in the Age Groups 10-14, 15-19, and 20-24 in the Major States of India for the year 1991.

Major states	Proportional Singles			SMAS (Years)	
	10-14	15-19	20-14	1991	1998-98
Andhra Pradesh	0.96	0.54	0.11	18.3	18.3
Bihar	0.93	0.44	0.07	17.5	18.8
Gujarat	0.98	0.77	0.18	19.9	21.1
Haryana	0.98	0.65	0.11	18.8	21.4
Himachal Pradesh	0.99	0.83	0.20	20.3	NA
Karnataka	0.97	0.73	0.24	20.1	20.1
Kerala	0.99	0.89	0.43	22.2	22.7
Madhya Pradesh	0.91	0.48	0.08	17.8	18.9
Maharashtra	0.98	0.72	0.17	19.7	21.3
Orissa	0.98	0.78	0.22	20.2	21.2
Punjab	0.98	0.88	0.29	21.0	22.1
Rajasthan	0.87	0.44	0.06	17.5	18.3
Tamil Nadu	0.98	0.82	0.28	20.9	21.7
Uttar Pradesh	0.93	0.53	0.08	18.0	21.5
West Bengal	0.98	0.66	0.19	19.7	19.6
All India	0.95	0.64	0.17	19.3	19.7

Source: (i) Census of India, 1991, part IV AC-Series, social and tables, vol.1 (table c-1)
(ii) IIPS, National Family Health Survey, Second Round (1998-99).

Nevertheless, the major states like Bihar, Madhya Pradesh, Rajasthan, and Uttar Pradesh in the Hindi belt of the north continued to reveal a higher incidence of child marriage up to the close of the century. Against this, the southern states of Kerala, Karnataka, and Tamil Nadu have occupied the lowest position in terms of the prevalence of child marriage throughout the period. In fact, early marriage, particularly child marriage, was not common in the whole of south India, except in Andhra Pradesh, even at the close of the nineteenth century. Table 2 presents 'proportion single' women in the age groups 10-14, 15-19, and 20-24 in the major states of India for the year 1991. Despite the fact that the incidence of child marriage has considerably declined in the country, in Rajasthan, 13 percent of the females in the age group 10–14 were returned as married. The states of Bihar, Madhya Pradesh, and Uttar Pradesh are only marginally behind Rajasthan in regard to the incidence of child marriage. In fact, the actual magnitude of child marriage would be much greater than what appears from the census. Remember that the census treats the population below 10 years of age as never married, irrespective of their marital status. Singulate mean age at marriage for females in these states also works out to be the lowest in the country. Remarkably, in all them, except Uttar Pradesh, the average age at marriage is well below the prescribed minimum. In Uttar Pradesh also, the average just equals the minimum. On the other extreme, Kerala and Andhra Pradesh, the average just equals the minimum. reports the average age of over 22 years. Thus, the gap in the mean age at marriage between the highest and the lowest is as much as 4.74 years. The other states reporting mean age at Madhya Pradesh Orissa are Punjab, Tamil Nadu, Himachal Pradesh, Kerala, Orissa, and Karnataka. It should be noted that Kerala literacy had a somewhat higher age at marriage even when the literacy rate among females was not that high. Keral the pre-puberty cutoff in the country, which had crossed pre-puberty cutoff level, i.e., 15 years, as early as in the nineteenth century The other southern states of Karnataka and Tamil Nadu achieved this distinction only in the middle of the twentieth century. Punjab was the leading state from the north to have achieved this level in the early decades of the twentieth century. Another example of rapid increase in the mean age at marriage comes from West Bengal. It may be recalled that the state ranked very high in terms of the prevalence of child marriage till the early decades of the twentieth century. The state is found to have recorded an impressive increase in the age at marriage in the post-independence period, particularly after 1961. Maharashtra and Gujarat, which experienced rapid urbanization and industrialization in the post-impendence period, also experienced a very quick rise in age at marriage.

India's population has been characterized by a marked contract in marital status between the urban and rural areas. Age at marriage is almost invariably lower in the rural areas than that in some of the early age groups, which is always lower in the former than in the latter. In the age group 10-14 years, over 5 percent of the females in rural areas were returned as 'currently married' as against only 2 percent in the urban areas. Similarly, 40.36 percent of females aged 15-19 years in the rural areas were already married at the time of the 1991 census, whereas in the urban areas the corresponding figure was as low as 21.32 percent. The social, cultural, and economic were as low as 21.32 percent. The social, cultural, and economic conditions in the rural areas in general favor an early marriage. Thus, singulate mean age at marriage has always been lower in rural areas.

Table 3: Rural-Urban Differentials in Marital Status of Females (Aged 10-20 years) in India, 1991

Age group	Percentage Distribution of Population					
	Never Married	Married	Windowed	Divorce/ Separated	Un-specified States	Total
Rural						
Areas						
10-14	94.59	5.30	0.06	0.03	0.02	100.00
15-19	58.50	40.68	0.20	0.31	0.01	100.00
20-24	12.60	86.07	0.64	0.68	0.01	100.00
Urban						
Areas	97.70	2.18	0.05	0.03	0.04	100.00
10-14	78.41	21.32	0.12	0.12	0.03	100.00
15-19	28.52	70.62	0.45	0.39	0.01	100.00
20-24						

Source: Census of India, 1991, part IV AC- Series, Social and Cultural Tables, vol.1 (Table c-1).

Census data indicate that the difference in age at marriage between rural and urban areas has declined since 1971. In the year 1971, the difference was of the extent of 2.5 years, which has come down to 1.1 years in 1991. Notably, rural areas have recorded a faster increase in age at marriage during the recent past. The difference in rural and urban age at marriage can be seen in all the states of India. It should be noted here that the states reporting a lower age at marriage than the national average also reveal a higher rural-urban differential. The regional variation in age pattern in India based on state-level data conceals more than what it reveals. In fact, the regional variation in the social and cultural attributes affecting marriage patterns is often as great within a stage as between the states. The coastal areas differ markedly from those in the interior peninsula. Even Uttar Pradesh, which reports one of the lowest average ages at marriage in the country, reveals significant variation from one part to another. The peninsular India, on average, excluding the Telangana region in Andhra Pradesh, reports a higher age at marriage. In the peninsula also, the coastal districts reveal a higher age at marriage than the interior uplands. The coastal districts of Kerala, Karnataka, and Maharashtra appear distinct on the map of the total 59 districts with a mean age at marriage above 21 years in the country as per the 1991 census (excluding Assam and Jammu & Kashmir); 33 districts are located along the coast in the west and south. The rest of the districts are located in the north and in the northeastern states. As noted earlier, Kerala ranks first in the country in terms of age at marriage. Within this state also, the southern parts report a higher age at marriage than their counterparts in the north. The adjoining districts of Tamil Nadu also reveal a very high age at marriage. In the north, the northern parts of Punjab and Himachal Pradesh, and the districts in Nagaland, Manipur, and Mizoram in the northeast report age at marriage above 21 years. As against this, in a vast area spread over much of the Bihar plains, eastern Uttar Pradesh, the northern parts of Madhya Pradesh, and eastern Rajasthan, the mean age at marriage is well below 18 years. It may be noted that of the total of 126 districts in the country, the mean age at marriage is below 18 years. It may be noted that of the total of 126 districts in the country with a mean age at marriage below the prescribed minimum level of 18 years, as many as 114 come from the four states of Bihar, Madhya Pradesh, Rajasthan, and Uttar Pradesh alone. The only other region with a very low age at marriage lies the central parts of the peninsular upland in the Telangana region of Andhra Pradesh and its adjoining districts in Maharashtra. In the peninsula, as one moves away from coastal areas, the mean age at marriage tends to decline. An exception to this, however, can be seen in somewhat higher ages at marriage in the eastern parts of Maharashtra in the Vidarbha region. In Orissa, high age at marriage extends over much of the upland plateaus. In Gujarat, the Katch and the Saurashtra peninsula report a higher age at marriage than their counterparts in the north and east. In Rajasthan also, the western parts appear conspicuous with a higher age at marriage than its eastern parts. Likewise, in Uttar Pradesh, which reports an average marriage age of 18 years at the aggregate level, a marked contrast can be observed between the western and eastern parts. In the western parts of the state, the mean age at marriage varies between 18 and 20 years in some of the districts exceeding 21 years while in the eastern region adjacent to the Bihar Plain it is well below 18 years.

Table 4: Age at marriage for Women in India

States	1992	2002	2012	2017	2022
Total	16.8	17.1	18.0	18.5	18.9
Andaman & Nicobar	NA	NA	NA	19.8	20.8
Andhra Pradesh	15.4	15.7	16.8	17.4	18.1
Arunachal Pradesh	18.1	18.2	18.7	19.4	20.5
Assam	17.2	18.0	18.7	19.1	19.4
Bihar	16.2	16.2	17.0	17.5	17.2
Chandigarh	NA	NA	NA	20.1	20.7
Chhattisgarh	NA	NA	NA	18.3	18.8
Dadar & N. Haveli	NA	NA	NA	18.6	18.2
Daman and Diu	NA	NA	NA	19.5	19.2
Goa	20.6	21.8	22.7	23.2	24.1

States	1992	2002	2012	2017	2022
Gujrat	18.1	17.9	18.5	19.0	19.5
Haryana	17.2	17.6	18.4	19.0	19.2
Himachal Pradesh	17.7	18.8	19.7	23.3	21.1
Jammu and Kashmir	17.9	18.4	20.0	21.2	23.5
Jharkhand	NA	NA	NA	17.7	18.0
Karnataka	16.7	17.3	18.4	19.0	18.9
Kerala	19.6	20.0	20.6	20.9	21.3
Lakshadweep	NA	NA	NA	21.2	22.3
Madhya Pradesh	16.0	16.2	17.2	17.8	18.4
Maharashtra	16.4	17.0	18.2	18.6	19.3
Manipur	20.0	20.6	21.7	22.2	22.3
Meghalaya	19.2	19.0	19.8	20.3	20.9
Mizoram	20.9	20.5	20.8	21.4	22.7
Nagaland	19.5	19.4	20.2	21.2	23.0
New Delhi	18.5	19.9	19.1	19.4	20.3
Odisha	16.7	17.4	18.4	19.1	19.4
Puducherry	NA	NA	NA	20.6	21.8
Punjab	18.9	19.3	19.9	20.6	20.6
Rajasthan	16.2	16.4	17.3	18.0	18.9
Sikkim	NA	19.0	19.7	20.2	20.7
Tamil Nadu	17.9	18.5	19.2	19.7	20.7
Telangana	NA	NA	NA	17.6	17.5
Tripura	19.3	18.0	18.6	18.9	18.6
Uttar Pradesh	16.7	16.4	17.6	18.3	18.9
Uttaranchal	NA	NA	NA	19.0	19.4
West Bengal	16.3	17.5	17.4	17.1	17.5

Source: globaldatalab.org

Interpretation of Age at Marriage for Women in India (1992-2022)

The data presented indicates the average age at marriage for women in different states and Union Territories in India from 1992 to 2022. This measure reflects the age at which women, on average, marry for the first time, offering insights into socio-economic development, educational attainment, and evolving social norms across regions. The age at marriage is a key demographic indicator that influences fertility rates and can affect women's health, economic participation, and empowerment.

The national average age at marriage for women has steadily increased over the period, from 16.8 years in 1992 to 18.9 years in 2022. This indicates a clear trend of delayed marriage for women, which is likely due to improvements in education, employment opportunities, and the rise of social norms favoring later marriages. Goa stands out with a consistently high and increasing age at marriage, from 20.6 years in 1992 to 24.1 years in 2022, reflecting higher levels of education and socio-economic development. Jammu and Kashmir shows a significant increase in the age at marriage, from 17.9 years in 1992 to 23.5 years in 2022, which may be influenced by changes in educational access and social attitudes toward marriage. Himachal Pradesh, Kerala, and Nagaland also demonstrate notable increases in age at marriage, with Nagaland rising from 19.5 years in 1992 to 23.0 years in 2022. Andhra Pradesh and Madhya Pradesh show a steady but modest increase in the average age at marriage. Andhra Pradesh rises from 15.4 years in 1992 to 18.1 years in 2022, while Madhya Pradesh moves from 16.0 years in 1992 to 18.4 years in 2022. This suggests some improvements in education and socio-economic development, though marriage ages remain lower compared to more developed regions. Uttar Pradesh shows an increase from 16.7 years in 1992 to 18.9 years in 2022, indicating a shift towards later marriages, but the age at marriage remains relatively lower compared to states like Goa and Kerala. Bihar, Rajasthan, and West Bengal have shown relatively slow increases in the age at marriage over the years. For instance, Bihar has had a modest increase from 16.2 years in 1992 to 17.2 years in 2022, indicating that early marriages are still prevalent in the state. West Bengal shows minimal increase, with the age at marriage staying relatively low at 17.5 years in 2022, suggesting slower changes in societal norms regarding marriage age. Andaman & Nicobar Islands, Chandigarh, Dadra & Nagar Haveli, Lakshadweep, and several other Union Territories show missing data for earlier years, making it difficult to track trends for these regions over the long term.

Social and Economic Factors

The increase in the age at marriage, particularly in more developed states like Goa, Kerala, and Jammu and Kashmir, reflects a trend toward greater educational attainment and career opportunities for women. As women delay marriage, they may gain greater independence, better access to health services, and improved economic prospects. In contrast, states with lower increases in the age at marriage, like Bihar and Rajasthan, may continue to face challenges related to child marriage, lower female educational attainment, and limited socio-economic empowerment for women. Delayed marriages are typically associated with delayed childbearing, which can lower fertility rates and contribute to a demographic transition, as women in higher-income states may have fewer children later in life. Conversely, in states with lower marriage ages, fertility rates may remain higher, contributing to more rapid population growth. As the age at marriage increases, policy interventions could focus on promoting higher education for women, improving health services, and offering financial incentives to women who choose to delay marriage for career development or personal growth. On the other hand, in states with lower marriage ages, efforts to raise awareness about the benefits of delaying marriage and improving access to education for girls are key to promoting more sustainable demographic changes. The data indicates significant regional disparities in the age at marriage for women in India, with some states showing substantial increases, while others have seen modest or slow changes. These trends reflect varying levels of development, education, and changing social norms across the country. Understanding these differences is critical for addressing the diverse challenges and opportunities related to women's empowerment, fertility rates, and population growth in India.

CHAPTER-7

Population Policies

India's Population Policy

While considering the population policy of India, it is necessary to concentrate on fertility as the single most important factor contributing to population change. As for international migration, the volume of such migration is not very large, and as for migration within the country, it is a constitutional privilege, and the question of any special policy in this regard does not arise. It, therefore, becomes imperative to study in detail India's policy related to fertility and mortality. This chapter will be devoted to these two aspects of the population policy of the country. Fertility-Influencing Policy Since India has the distinction of being the first country in the world to launch a nation-wide family planning program with full government support, it is of interest to study the historical evolution of her anti-natalist policy and to critically assess the programs undertaken in pursuance of this policy in the light of her achievements over the years. It is also necessary to take note of the shifts in the approach and the emphasis to identify the gaps in the policy and the programs. Such an analysis of India's anti-natalist population policy has to be undertaken within the broader framework of programs for socio-economic development, which India has undertaken from time to time. The Pre-Independence Period the British rulers in India showed little interest in developing a population policy for the country. The absence of support from the British rulers to the birth control movement was firstly because in their own homeland the birth control issue was itself controversial and, secondly, because the general policy of the British was to keep away from any measures that would be considered by the Indians as an intrusion their own traditions, customs, values, and beliefs. A section of the intellectual elite among the Indians showed some concern about the population issue during the period between the two World Wars, despite the fact that the preoccupation of the general population was the density of population rather than the rate of growth, for high rates of mortality as well as of fertility did not result in alarmingly high growth rates. As per 1931 census, Population has increase rapidly than that of earlier census. Neo-Malthusians believed that a smaller population would improve living conditions for the masses. Following Malthus's ideas, they argued that rapid population growth would negate economic gains and advocated for a population policy promoting birth control.

Several key developments took place between 1916 and 1947 to attain the independence in. In 1916, Pyare Kishen Wattal published his book, The Population Problem Advocated in India, in which he advocated family planning. In 1925, Raghunath Dhondo Karve, a professor of mathematics in a college run by Christian missionaries, opened the first birth control center in Bombay and faced dismissal at the hands of his orthodox employers. In the same year, Rabindranath Tagore wrote to Margaret Sanger. "I am of the opinion that the birth control movement is a great movement not only because it will save women from enforced and undesirable maternity but also because it will help the cause of peace by lessening the number of surplus populations of a country scrambling for food and pace outside its own rightful limits. In a hunger struggling country like India, bringing more children into existence than can be cared for is seen as a cruel crime. In 1930, Mysore opened the world's first government birth control clinic. Madras University proposed education on conception control in 1931, and clinics were approved in 1932 by the All-India Women's Conference. In 1935, the Indian National Congress established a National Planning Committee led by Jawaharlal Nehru. As the recommendations of this committee are of historic subsistence, we agree with the view that the size of the Indian population is a basic issue in national economic planning, in so far as its unrestricted increase in proportion to means of subsistence affects adversely the standard of living and tends to defeat many social ads that have been fundamentally caused by the lack of all-round coordinated economic development. While population and limiting excessive population pressure are necessary, the basic solution of the present disparity between population and standard of living lies in the economic progress of the country on a comprehensive and planned basis. In the interest of the social economy, family happiness and national planning, family planning, and the limitation of children are essential, and the state should adopt a policy to encourage these. It is desirable to lay stress on self-control as well as to spread knowledge of cheap and safe methods of birth control. Birth control clinics should be established and other necessary measures taken on this behalf to prevent the use or advertisement of harmful methods. We consider that a gradual raising of the marriage age and discouragement of polygamy are desirable in the interests of the size of the family. A eugenic program should include the sterilization of persons suffering from transmissible diseases of a serious nature, such as insanity or epilepsy. The maintenance of vital statistics and the carrying out of periodic demographic surveys on comprehensive lines are necessary, and appropriate machinery should be devised for the purpose." On December 1, 1935, the Society for the Study and Promotion of Family Hygiene was founded, with Lady Cowasji Jehangir as president. Dr. A. P. Pillai conducted training courses in birth control. By 1939, clinics were established in Uttar Pradesh and Ujjain, and in 1940, P.N. Sapru successfully proposed more clinics in the Council of States. Around the same time, Shrimati Renu Datta toured the country extensively with the objective of organizing birth control campaigns on behalf of the Family Planning Society, incorporating the Birth Control Clinic of the Bhagini Samaj in Bombay.

In the midst of all this support for population control and family planning, a different note was sounded' by the Famine Enquiry Commission of 1943, called the Woodhead Commission, which stated: "At the present time, a deliberate state policy with the objective of encouraging the practice of birth control among the mass of the population is impracticable. Economic development is the reason behind declining birth rate. In 1945 under the chairmanship of Sir Joseph Bhore, Health Survey and Development Committee set up by the Government of India again the committee recommended the birth control services should be provided for the promotion of the health of mothers and children. The impact of Gandhiji's view on birth control also needs to be considered because he exercised a strong influence over the Indian masses, not only in the political field but in all areas of human life. Though Gandhiji accepted the need for birth control, it was advocated on ethical and moral grounds, for he maintained that would improve the moral fiber and the physical health of the individual. On the same ethical and moral grounds, he was against any artificial means of contraception, but he lent his full support to bramacharya, or abstinence. He declared, "There can be no two opinions about the necessity of birth control. But the self-control is the best practice to control birth. He, however, conceded that another method, which had an appeal for him, was the rhythm method, which was based on avoidance of sexual union during "unsafe" periods, that is, when conception was most likely to occur. Gandhiji's views on whether or not India was overpopulated are also worth noting. He wrote, "If all labored for their bread and no more, then there would be enough food and leisure for all, and there would be no cry of overviews on the enslavement of Indians by the British. This also made him advise his people, "It is our duty at the present movement to suspend bringing forth heirs to our slavery."

Some national leaders opposed the birth control movement, arguing that India's issues stemmed from foreign domination, not from Indian childbearing practices. Before independence, birth control was a contentious issue primarily among intellectuals, with its practice limited to a westernized urban minority. The pressure from the intellectuals that the government formulate a policy for disseminating information on birth control and for encouraging its practice was, however, mounting. As Gunnar Myrdal puts it, "The intellectual groundwork had been laid when the colonial era came to an end; a rational policy could be pursued in the era of independence." Family planning in independent India (1951-1996) The Five-Year Plans: A review of the eight development plans of the country indicates that family planning as a measure of population control has been given a high priority in each successive plan, which is also indicative of the growing importance accorded to the program. Beginning with the first five-year plan (1951–1996), the population issue has engaged the attention of the Planning Commission. The Draft Outline of the First Plan, published in July 1951, contained a section on "Population Pressure: Its Bearing on Development," which recognized that India has a population problem. "The increasing pressure of population in India is already so high that a reduction in the rate of growth must be regarded as a major desideratum." The Second Five-Year Plan (1951–1961) pointed out that the rate of population increase was one of the key factors in development and underscored. The result of the high population growth is negative growth of economic advance and living standards per capita. Given the overall shortage of land and/or capital equipment relative to population as in India, the conclusion is inescapable that an effective curb on population growth is an important condition for rapid improvements in income and in levels of living." It is important to note that the Planning Commission has never considered a population control program as an alternative to socio-economic development. While recognizing that with improved public health programs and lower death rates, the population pressure was likely to increase, it accepted the need for curbing the birth rates. "This highlights the need for a large and active program aimed at retraining population growth, even as it reinforces the case for a massive developmental effort." The Third Five-Year Plan (1961-1966), while considering population control in the context of long-term development, stated: "The objective of stabilizing the growth of population over a reasonable period must therefore be at the center of planned development." The Fourth Five-Year Plan (1969-1974) emphasized that controlling population growth is essential for achieving economic development and social change in India, linking family planning to the quest for equality and dignity.

The Draft Fifth Five-Year Plan (1974–1979) included family planning in the context of the National Minimum Needs Programme. It accorded a high priority to population control and based the demographic projections for the Plan period on certain assumptions regarding a reduced growth rate, a birth rate declining much faster than the death rate, and an effective family planning program.

The final version of the Fifth Plan was published only in 1976, and it incorporated the National Population Policy announced in April 1976. It reiterated: "The policy envisages a series of fundamental measures, including raising the age of marriage, female education, spreading population values and the small family norm, strengthening research in reproductive biology and contraception incentives for individuals, groups, and communities, and permitting state legislatures to enact legislation for compulsory sterilization." A target was set for a birth rate of 25 per thousand and a 14% population growth rate by the end of the Sixth Plan period, with expectations to achieve these goals. The government that was voted into power in March 1977 made it explicitly clear that it was firmly committed to the program of population in the program did not arise. Some of the measures announced in the National Population Policy in 1976, especially those relating to compulsion, were dropped, while some others were retained. The government that came in power was also committed to the program of population control and also ruled out any element of compulsion in the program. The Sixth Five-Year Plan (1980-1985) earmarked Rs. 10,100 million for the family planning program. The long-term demographic goal recommended by the Working Group on Population Policy appointed by the Planning Commission was approved by the National Development Council. It involves reducing the net reproduction rate (NRR) to one by 1996 for the country as a whole and by 2001 in the states. The implications of this long-term demographic goal are as follows: The average family size should be declined 4.2 children to 2.3 children. The birth rate per thousand population would be reduced from the level of 33 in 1978 to 21. The death rate per thousand population would be reduced from about 14 in 1978 to 9, and the infant mortality rate would be reduced from 129 to 60 or less. As against 22 percent of the eligible couples protected with family planning, 60 percent would be protected. The population of India will be around 900 million by the turn of the century and will stabilize at 1,200 million by the year 2050 A.D." In the Seventh Five-Year Plan (1985–90), Rs. 3,256 crores were allocated for family welfare, with the goal of achieving a net reproduction rate of unity projected for 2006-2011, delayed from the original target of 2000 as envisaged in the Sixth Five-Year Plan. The goals set for 1990 included achieving a birth rate of 25 per thousand, reducing the population growth rate, and improving family welfare services. Effective couple protection rate: 42 percent Crude birth rate: 29.1 Crude death rate: 10.4 Infant mortality rate: 90.0 In order to reach an effective couple protection rate of 42 percent, 31 million sterilizations and 21.25 million IUD insertions would have had to be performed by 1990, and during the terminal year of the Seventh Plan, 14.5 million users of conventional contraceptives/oral contraceptives would have had to be enlisted. The targets to be reached with regard to different methods in the Seventh Plan are given in Table 14.2. In the Eighth Plan (1992-97), the total outlay for the Family Welfare Programme was Rs. 6500 = Rs. 00 crores. Keeping in view the level of achievement, the long-term demographic goals for the country have been revised. It has been stated in the Eight Five Year Document that the net reproduction rate of one would now be achievable only in the period 2011-16 A.D. and not by the year 2000 A.D. as envisaged in the National Health Policy declared in 1983. Accordingly, the goals of the crude birth rate, crude death rate, and natural population growth rate to be achieved by the end of the Eight Plan Period (i.e., 1997) were also revised. The revised goals to be achieved by 1997 and 2000 A.D., respectively, are as follows: crude birth rate: 26 and 21; crude death rate: 9 and 9; The goal for the effective couple protection rate by 1997 was 56 percent, and by the end of 2000 A.D. it continued to be 60 percent. The latest report of the Technical Group on Population Projections constituted by the Planning Commission has estimated that the replacement level of NRR-1 is achievable by 2026 A.D. and beyond.

The Family Planning Programs

In this section, the family planning program will be reviewed with reference to the following points: (1) organizational structure; (2) approach to family planning program implementation; (3) family planning methods; (4) achievements. Organizational Structure: The family planning program, since its very inception in 1952, has been the responsibility of the Ministry of Health. It is a centrally sponsored and financed program implemented by the States. It began very modestly indeed, with the creation of a Family-Planning Cell I in the Planning and Development Section of the Directorate-General of Health Services in April 1952. The first Officer on Special Duty to be in charge of the program was appointed in March 1956; his designation was later changed to Director of Family Planning. At the state level, family planning boards were set up at the center as well as in the states during the second five-year plan. In 1966, several important developments concerning the family planning program took place. A full-fledged Department of Family Planning was established within the Ministry, which was redesignated as the Ministry of Cabinet rank was placed in its charge. A Cabinet Committee of Family Planning, initially headed by the Prime Minister and later by the finance minister, was constituted at the central level. The function of the Cabinet Committee was to give proper direction and policy clearances and to regularly review the progress of the program. In addition to the finance minister, this Cabinet Committee consisted of the Minister of Information Broadcasting, the Minister for Food and Agriculture, the Minister for Labour and Employment, the Deputy Chairman of the Planning Commission, and the Minister of Health and Family Planning. Similar committees were constituted at the state level and were headed by the Chief Minister of the State. The Central Family Planning Council, with the Minister of Health and Family Planning as its Chairman, was set up in 1966. The council has a Standing Committee. Similar councils were set up in the States as well. These councils are mainly meant to review the progress of the family planning program and to make policy decisions. They consist of experts as well as representatives of the government, the meetings of the Central Council of Health and the Central Council of Family Welfare, as it is now known, are held jointly., voluntary organizations, industries, etc. At present, the meetings of the Central Council of Health and the Central Council of Family Welfare, as it is now known, are held jointly. The First United Nations Advisory Mission, which visited India in 1955 to make recommendations for improvement in the family planning program, had observed, "The Mission feels that the present Directorate for Family Planning is in the administrative hierarchy and needs internal reorganization." In 1966, with the appointment of a Commissioner of Family Planning with the rank of Director-General, this lacuna was filled to a certain extent. At the state level, the State Family Planning Bureau is headed by the Joint Director, Family Planning and Maternal and Child Health, and an officer of the rank of a Secretary/Deputy Secretary head the cell in the Secretariat. Each district is entitled to its own Family Planning Bureau with a Class I officer in charge. The District Bureau has divisions for administration, education and information, field operation, and evaluation. Mobile units for IUD and sterilization are also attached to the District Bureau. The headquarters of the rural family planning organization is the Primary Health Centre, with sub-centers attached to it. For the urban family welfare planning centers, different patterns of staffing are provided, according to the size of the population to be covered. Each city with a population above 2 lakhs is entitled to its own Family Planning Bureau. All cities with a population of more than 10 lakhs are allowed the pattern of the District Family Planning Bureau. Different patterns are laid down for the family planning units of public sector undertakings and those of the Ministry of Railways, Posts, and Telegraphs Department and the Ministry of Defense, depending on the size of the population to be covered.

In 1977, when the Janata Government came into power, the family planning program was renamed the family welfare program to make it more acceptable and to denote that it had a wider base so that it could also include various measures of family welfare for the improvement of the quality of human life. This review of the development of the organizational structure of the family planning program is indicative of the rapidity with which the program gained in importance. Another point that emerges in this review is that this program has always been considered to be the responsibility of the Ministry of Health. Some critics of the family planning program look upon this very fact as the reason for its rather poor performance over the years. "As the Health Ministry is one of the weakest and politically unimportant ministries both at the Center and in the States, family planning is relegated to an unimportant position, in spite of pronouncements to the contrary." The solution to this problem may either lie in the setting up of a separate ministry for the implementation of the program. The first U.N. Advisory Mission carefully considered both these proposals. "It rejected both of these proposals because it felt that as long as the most effective methods of contraception required medical supervision and cooperation and were quite rightly closely identified with maternal and child welfare, family planning should remain the responsibility of the Ministry of Health." In recent times, there has been one school of thought that favors the setting up of an Indian Commission on Population "to recommend policies and programs with respect to all aspects of India's population." The New Delhi Declaration of Parliamentarians issued at the First National Conference of the Indian Association of Parliamentarians for Problems of Population and Development has in May 1981 also recommended, "A National Population Commission should be set up immediately to review implementation of population policies and programs, report progress, and provide concrete suggestions to Parliament." Approach to Family Planning Programme Implementation: Since its inception in 1952, the family planning program has undergone several revisions that, despite the criticism they have called forth, nevertheless show that those who administer the program are willing to learn from experience. It is obvious that rigidity in the implementation of the program would not have been the correct stance to adopt. The Clinic Approach: The family planning program in India started with a very cautious approach. The first five-year plan emphasized field research with a view to identifying values, norms, customs, and beliefs concerning childbearing. The natural method of family planning (the Rhythm Method) was considered to be the most appropriate for the Indian masses, and this method was propagated. It may be pointed out here that India was the first country in the world to adopt family planning as a nation-wide program and that there was no model that she could follow. The only model that was available for this purpose was the one used by the Planned

Parenthood Organizations in the West, which set up family planning clinics, those in need of family planning were expected to fully utilize the available facilities. The limitations of such an approach may well be imagined. It assumes that individuals needing family planning will visit clinics without hesitation. Such an assumption, however, ruled out the need to reach out to people to educate them about the need for family planning. Moreover, the planned parenthood model leaned heavily in favor of services for women rather than for couples. It also leaned heavily on medical personnel, which was and is rather scarce in India. This "clinic approach," as it was called, "could be expected to reach only a relatively small fraction of the people and could not be expected to make much impact on birth rates." The approach was, therefore, not considered suitable for a developing country like India. Hence, the need for the adoption of the "extension approach" to family planning was soon realized.

The Extension Approach

In 1963, in his report of historic importance, Lt. Col. B.L. Raina highlighted the need to strengthen the extension approach, which involves the adoption of an educational approach to bring about changes in the knowledge, attitudes, and behavior of the people in regard to family planning; it also involves the acceptance of the principle that "the power, inherent in a group itself to bring about changes in deeply rooted practices among the members of the group, is greater than the influence of individual instruction by outsiders." In the extension approach, therefore, influential formal and informal leaders in different sub-groups of the population are first identified and then encouraged to gain knowledge and to take interest in popularizing the acceptance of the small family size norm among their own group. This approach advocates for collaborating with the community instead of imposing solutions from the outside. The extension approach, as outlined in the Report of the Director of Family Planning for 1962–63, envisages certain operational goals for the family planning program with a view to reducing the birth rate as expeditiously as possible to 25 per thousand populations. These operational goals should "create, for 90 percent of the married adult population of India, the three basic conditions needed for accelerating the adoption of family planning by couples: group acceptance, knowledge about family planning, and easy availability of supplies and services." It is necessary to elaborate on these three conditions. The emphasis on the need for group acceptance of family planning recognized the fact that individuals do not live in a vacuum but are greatly influenced by the group in which they live. It is difficult for individuals to accept new forms of behavior unless they are acceptable to their group as a whole. If family planning is to be adopted by individuals, expect them to adopt family planning by becoming "rebels" or "deviants. While the first condition for the acceptance of family planning is socio-psychological and may be considered to be somewhat difficult to define, the other two are more practical. It is obvious that if a couple has to practice family planning, knowledge about family planning is essential—knowledge that it is possible to prevent conception, knowledge about the various methods of contraception so that the most suitable method may be chosen, and knowledge about where such methods are available. The third condition, "easy availability of supplies and services," has also to be fulfilled if couples are to practice family planning. If strong enough, a couple would seek out the sources of supplies and services even if they are not easily available. But such strong motivation is not always forthcoming. It is therefore better to facilitate the practice of family planning by removing both physical and psychological barriers associated with the usual sources of services for that purpose. If a family is planning by removing both physical and psychological barriers associated with the usual sources of services for the purpose. If a family planning center is at a considerable distance and the clinical atmosphere is quite imposing, both of these factors may act as barriers to the acceptance of the family planning program. The Integrated Approach: The "integrated approach" to family planning, which has assumed importance today, is not of recent origin. Even in the Second Five-Year Plan period, it was clarified that the "Family Planning Service is likely to succeed if the clinics are associated with maternity and child health work or with centers that provide medical aid and welfare services." The principle of integration of family planning services with maternal and child health services has thus been accepted almost since the beginning of the planning program. An explicit reference to this aspect is found in the Draft Fifth Five-Year Plan. "According to the approach formulated for the Fifth Five-Year Plan, the Family Planning Programme will form an integral part of Maternal and Child Health (M.C.H.) and Nutrition Services. With this change in strategy, it is expected that the program will find a wider acceptance among the people." The rationale for such an integration is that when infant and child mortality rates are high, as in India, parents cannot be expected to limit the size of their families unless they have some confidence that the children they already have will survive to adulthood. Such confidence can be created only by providing preventive and curative medical services for children. It is also obvious that while providing such services, health workers may approach mothers for family planning with greater credibility. At present, the special schemes under the family planning program include the following maternal and child health programs: (i) Immunization of infants and preschool-age children against D.P.T.; (ii) Immunization of expectant mothers against tetanus; (iii) Prophylaxis against nutritional anemia among mothers and children; and (iv) Prophylaxis against blindness caused in children by vitamin "A" deficiency. The emphasis laid on the provision of maternal and child health services in the family planning program may be highlighted by pointing out that one-fourth of the entire outlay for family planning in the Sixth Plan is meant for maternal and child health services.

With the appointment of multi-purpose workers, a new concept in the delivery of health nutrition and family planning services has been introduced. It was pointed out by the Committee on Multipurpose Workers under the Health and Family Planning Programme, which was appointed by the Government of India in 1973, that while many health administrators were of the view that the staff of the primary health centers and sub-centers were not able to adequately meet the health and family planning needs of the population to be covered, community leaders complained that rural folk were not happy about the fact that too many workers were visiting their homes for individual health programs and that the present health and family planning workers were not in a position to provide remedies for even simple ailments like headaches, cuts, and burns. The health and family planning workers endorsed these views and added that their acceptability would increase if the expectations of the villagers were fulfilled. The entire country was expected to be covered by the Multi-purpose Workers Scheme during the Fifth Five-Year Plan Period. Under this scheme, comprehensive health and family planning care is provided through a team of two workers, one male and one female, at the sub-center level, with one sub-center for a population of 5,000. These multi-purpose workers are responsible for providing first aid and treatment for minor ailments in addition to health and family planning care and nutrition education. "It is envisaged that, under the scheme integration and health and family planning services and nutrition, education will take place from the sub-center through the P.H.C. (Primary Health Centre) and district levels to the States."

In October 1977, the Community Health Workers' Scheme was introduced in selected primary health centers. This scheme has the objective of providing preventive and promotional services of health and family planning through a trained community health worker selected but the community itself for every village or a population of one thousand. This worker is also responsible for treating minor ailments and for providing referral services. The designation of this trained person was later changed to "Community Health Volunteer" to emphasize the fact that he was not a government employee and to highlight the principle of community participation and involvement. Recently, it has been decided that this person will be known as the village health guide. The salient features of the revised scheme are that females are to be selected as health guides and should preferably be 30 years of age or older and should be residing in the village permanently. Male health guides are to be selected only if females are not available, and preference should be given to ex-servicemen, freedom fighters, or persons known for their social service. As of 31-3-1988, there were about 4.01 lakh such trained persons in the country. During the Seventh Plan period (1985–90), it was proposed to give priority to the training and placement of village health guides. Another aspect of the integrated approach is the postpartum program, which is restricted only to hospitals, where women are enrolled either for delivery or for abortion. The all-India Hospital Post-Partum Programme was initiated in January 1970. By the end of 1974–75, there were 255 such centers in the country. (By the end of 1975-76, this number went up to 324.) By the end of 1979-80, this program covered 524 hospitals in the country, but these hospitals were either at the district level or above that level. In 1980-81, it was, for the first time, extended to 50 sub-district hospitals. During the sixth plan, the program was to be extended to 300 subdistrict hospitals. As of March 1989, 554 postpartum centers have been sanctioned in medical colleges, district hospitals, and maternity hospitals. Another 400 centers have been sanctioned for sub-divisional hospitals. More sub-divisional-level post-partum centers were to be established during the Seventh Plan period." The rationale for this program is that when women go to hospitals for maternity care, they are in a positive frame of mind to receive family planning and education, for the need for family planning is uppermost in their minds at that time. In fact, Zatuchni, Director of the International Post-Partum Family Planning Programme of the Population Council, New York, has listed fourteen advantages of such an approach, some of the more important of which stem from the very nature of a hospital, which makes it a respected institution with good resources capable of providing continuing service to its patients. The wide sphere of influence of hospital patients who return to their families is also mentioned as an advantage of this approach. An extension of the Integrated Approach was the introduction of the Child Survival and Safe Motherhood (CSSM) Programme in 1992-93. Details of this program are given in the Section on Mortality Influencing Policies of this chapter.

The Camp Approach

The idea of holding mass vasectomy caps was not entirely new when it was reintroduced in a big way in 1971-72. As early as 1961, the first vasectomy camp in the world was organized by the Government of Maharashtra, where 1,400 men were sterilized in 3 days. In November-December 1970, a massive vasectomy camp was held in the Ernakulum District of Kerala, where a total of 15,005 vasectomies were performed over a period of one month. This performance was repeated in July 1971 on a much larger scale, when 63,418 vasectomies were performed in a one-month period. Some of the salient features of these camps, which contributed to their spectacular success, have been identified by the moving spirit behind these camps, the Collector of Ernakulum District. These are: (i) Inter-departmental Cooperation: The District Collector successfully facilitated collaboration among various departments, enhancing resource sharing and support for the initiative; (ii) Festival Atmosphere: The camps created a welcoming and celebratory environment, encouraging community participation and reducing stigma around the issues being addressed.; (iii) The support from the representatives of the people as well as from industries that the organizer were able to muster; (iv) The special precautions taken while making technical arrangements in order to minimize infection and other complications following surgery; (v) Increased incentives to acceptors in addition to special prizes; and (vi) The employment of good management techniques for the organization of these massive camps.38 Encouraged by its success in Ernakulam, the Department of Family Planning allowed the States to organize such massive camps and provided additional support. The result was that the majority of vasectomies came to be performed in such camps. Of the 2.19 million vasectomies in 1971–72, about 61 percent were performed in mass vasectomy camps. In 1972-73, of the 3.12 million vasectomies, 83 percent were performed in such camps. This approach, however, has some limitations. In the first place, the setting up of a vasectomy camp involves tremendous organization and inter-departmental cooperation and coordination, which is often possible only at the cost of other developmental programs. It disrupts the normal activities of several departments and leads to lethargy in the implementation of the family planning program following concentrated activity in vasectomy camps. Even the slightest negligence on the part of doctors could have disastrous results, which would ultimately be counterproductive, as happened in Gorakhpur, where eleven men died of tetanus following vasectomy in a vasectomy camp. When, in 1973-74, the family planning budget was reduced by Rs. 6 crores, the exact amount earmarked for vasectomy camps, it was evident that the camp approach was on the way out. The report of the Minister of Health and Family Planning to Parliament made this point very clear. "While the number of vasectomy operations did touch a record level, it has been felt that, in a larger perspective, such camps might become counterproductive. It was, therefore, decided not to continue this strategy on a regular basis but to make an optimum utilization of available resources to strengthen the normal program, even at the risk of a decrease in the number of acceptors in the short term. As expected, the figures for vasectomy for the year 1973-74 dropped to 9.42 lakhs as against 3.12 million in 1972-73 and 2.12 million in 1971-72. Some states continued to organize camps for various family planning methods on a small scale, and these camps were known as mini-camps.

The Reproductive and Child Health Approach

The Population Policy of India underwent a major shift with the Programme of Action Adopted at the International Conference on Population and Development (ICPD) held in Cairo in September 1994. The Programme of Action linked the population issue directly to development, emphasizing its role in sustainable progress. It was recommended that instead of population control programs focusing on demographic targets, the emphasis should be on the needs of individuals. Population Policies, in the section on Role of the United Nations System in the field of population. It may be reiterated that at the ICPD, the nations of the world agreed that governments should give special attention to the education of girls, the health of women, the survival of infants and young children, and in general, the empowerment of women. It was also recommended that comprehensive reproductive health services should be provided to determine freely and responsibly the number and spacing of their children. Thus, the ICPD Programme of Action endorsed the concepts of reproductive and sexual health and rights and emphasized the need for providing services to achieve these goals. It is against this background that the Population Policy of India has to be studied, with particular reference to reproductive and sexual health rights and the provision of services required for meeting those goals. It is against this background that the Population Policy of India has to e studied, with particular reference to reproductive and sexual health ad rights and the provision of services required for meeting those goals.

Reproductive Health

Reproductive health has been defined by the World Health Organization as "a state of complete physical, mental, and social well-being and not merely the absence of disease." This definition of reproductive health implies two conditions. The first condition is that people are able to have a satisfying and safe sexual life, and the second condition is that they have the capacity to reproduce and the freedom to decide it. The second condition implies that men and women have the right to be informed and to have access to safe, effective, affordable, and acceptable methods of family planning by choice for regulation of fertility, which are not opposing the law of the land. It is also implied that people have the right to access appropriate health care services that would enable women to go through pregnancy and childbirth safely and provide couples with good chance of have a healthy infant. As one of the nation's adopting the ICPD Programme of Action, India took steps to incorporate this program in the national context. The joint mission of the Government of India and the World Bank set up in November 1994 recommended the Reproductive and Child Health Approach in its report submitted in 1995. The Government of India then decided to adopt the reproductive and child health approach to the Family Welfare Programme. The package of essential reproductive and child health services includes the following: (1) services for the prevention and management of unwanted pregnancy; (2) the promotion of safe motherhood and child survival; (3) nutritional services for vulnerable groups; (4) services for the prevention and management of reproductive tract infections and sexually transmitted infections; and (5) reproductive health services for adolescents. While services for the prevention and management of unwanted pregnancy include family planning and abortion-related services, services for the treatment of infertility are also included in the package of essential reproductive and child health services, as the fear of sterility and delayed childbearing are major barriers to contraceptive acceptance in the Indian cultural context. Services for safe motherhood and child survival have been an integral part of the Indian Family Welfare Programme for several years and continue to be so in the RCH Approach. As for services for reproductive tract infections, it is known that such infections are widely prevalent in India and demand the attention they deserve in the RCH Approach. With the advent of the HIV/AIDS pandemic, the inclusion of services for sexually transmitted diseases is fully justified. The adolescent group (10–19 years) has so far been neglected in all health services. Reproductive health services for adolescents would fill this gap to help them not only to prevent and manage reproductive health problems but also to deal with their own sexuality in a responsible manner. The RCH Approach not only suggests a package of integrated services but also an emphasis on individual needs rather than on demographic goals, which were often attempted to be reached through family planning method-specific targets, even using coercive means. The RCH Approach also implies providing client-centered quality services. It is obvious that all the services included in the RCH package cannot be introduced simultaneously throughout the country. It is necessary to prioritize and develop a phased approach so that health interventions can be added in an incremental way as resources become available and the implementers develop the necessary skills. In the meantime, the criteria suggested by Pachauri for prioritizing services to be included in the package could be as follows: (1) levels of fertility and mortality; (2) disease burden; (3) cost-effectiveness of available health interventions; and (4) the capacity of the health system to deliver health systems. In keeping with the spirit of the ICPD Programme of Action, the GOI did not fix contraceptive targets, on a trial basis, for Kerala and Tamil Nadu during 1995-96, as well as one or two districts in the other states. On February 1, 1996, the method-specific target approach was withdrawn from the entire country. By then, the limitations of fixing numerical method-specific targets have become evident. The pressure on the staff for fulfilling these targets was immense. The targets were allotted to individual states, which then allotted these targets to various districts as well as to individual health workers. This approach led to overreporting of performance by the staff and sometimes use of coercive methods. The quality of the services provided also suffered because of the undue emphasis on targets.

Highlights on National Population Policy, 2000 of India the National Population Policy (NPP) 2000 provides a policy framework for achieving goals and prioritizing strategies during the next decade to meet the reproductive and child health needs of the people of India, along with the target to achieve the net replacement levels (total fertility rate of 2.1 live births per woman). It aims for a stable population by 2045. The national socio-demographic goals aimed for achievement by 2010 included: To address unmet needs in basic reproductive health, the focus was on improving access to contraception, enhancing child health services, and strengthening healthcare infrastructure To make school education up to age 14 free and compulsory and reduce dropouts at primary and secondary school levels to below 20 percent for both boys and girls. To reduce the infant mortality rate to below 30 per 1,000 live births. To reduce the maternal mortality rate to below 100 per 100,000 live births. Universal immunization should be achieved for children against all vaccine-preventable diseases. Promote delayed marriages for girls, ideally after age 20. Ensure universal access to information, counseling, and a variety of contraception options for fertility regulation. To achieve 80 percent institutional deliveries and 100 percent deliveries by trained persons. To achieve 100 percent registration of births, deaths, marriages, and pregnancies. Prevent and control communicable diseases, particularly AIDS and sexually transmitted infections (STIs). To vigorously promote the small family norm. Integrate Indian Systems of Medicine (ISM) into reproductive and child health services to better reach households. Unlike China, India has not adopted a coerced one-child population control policy. It has been kept totally voluntary. For achieving the goals of the family welfare program, accredited social health activists have been appointed. The policy document projected that full implementation of NPP 2000 would result in India's population reaching 1,107 million by 2010, reducing it by over 55 million if the Total Fertility Rate (TFR) reached replacement level. The document stated a special strategic theme for underserved populations—slum populations, tribal communities, displaced migrant populations, and adolescents. Not only this, NPP 2000 had identified a separate strategic theme for the elderly for their health care and support. Ironically enough, the family planning (new welfare) program failed to deliver desired results at least until recently. Much of the failure is attributed to a strong preference for male children and widespread ignorance about birth control techniques among rural populations. Despite all efforts, there is little evidence of our capacity to arrest population explosions. An estimate suggests that about 8,000 people are added to it every day. According to the Human Development Report (2011), India ranked 134th out of 187 countries on the Human Development Index (HDI), which measures health (life expectancy), education, and standard of living.

Population and Sustainable Development Goals

Population growth significantly influences global development and resource use, with the world population exceeding 7.8 billion by 2020, bringing both challenges and opportunities. The United Nations, recognizing the intricate ties between population growth and global challenges, incorporated it indirectly into the Sustainable Development Goals (SDG) in 2015. The Sustainable Development Goals (SDGs) are 17 interconnected objectives designed to eliminate poverty, protect the planet, and promote prosperity for everyone by 2030. These goals, though not directly mentioning population growth, are profoundly influenced by it. Here's how: Rapid population growth increases pressure on limited resources, particularly in areas where they are already scarce, impacting SDG 2 (zero hunger) and SDG 6 (clean water and sanitation). A growing population demands more food, which can lead to unsustainable agricultural practices, while the rising need for freshwater strains reserves and may cause conflicts over access to these essential resources. The increase in population is driving significant migration to urban areas, with projections indicating that by 2050, two-thirds of the global population will live in cities. This urbanization poses challenges for achieving SDG 11 (Sustainable Cities and Communities). As the population grows, the demand for affordable, quality healthcare rises, directly impacting SDG 3 (good health and well-being). Overpopulated areas often struggle with disease outbreaks, insufficient medical facilities, and a shortage of healthcare professionals, complicating efforts to achieve this goal. Furthermore, the expanding global population also requires accessible quality education for everyone, aligning with SDG 4 (Quality Education). Moreover, the growing global population highlights the need for quality education for everyone, which aligns with the goals of SDG 4 (Quality Education). A larger population leads to increased waste, higher energy demands, and greater carbon emissions. This straight affects SDG 13 (Climate Action), SDG 14 (Life Below Water), and SDG 15 (Life on Land). Higher populations can strengthen deforestation, lead to improved greenhouse gas emissions, and expedite the degradation of marine and terrestrial ecosystems. However, population growth not only is a challenge; but also, is an opportunity. A growing population lead to larger workforce, higher number of innovative minds, and potentially a market for goods and services. Countries can harness this 'demographic dividend' for economic growth if they can sufficiently invest in the health, education, and employability of their young people. Role of sustainable development goals 2017 in population challenges The Sustainable Development Goals (SDGs), established in 2015 and effective until 2030, provide a global framework aimed at addressing various challenges, including those related to population dynamics.

Here's how the SDGs contribute to managing population challenges: Goal 1 is No Poverty Impact: To manage population growth, reducing poverty is critical. Lower poverty rates often correlate with improved health and education, leading to lower birth rates. Goal 3 is Good Health and Well-Being Impact: Ensuring access to health services, including reproductive health, directly influences population growth. Improved maternal and Child health outcomes lead to lower mortality rates and informed family planning. Goal 4 is Quality Education Impact: Education, especially for women and girls, empowers individuals to make informed decisions about family size and reproductive health. Higher Education levels often affect the fertility rates positively; the higher the education of the mother lower is the fertility rate observed. Goal 5 is Gender Equity Impact: Promoting gender Equality and women's empowerment are vital for population management. Women who have equal access to resources and decision-making tend to have fewer children and improve community health and economic outcomes. Goal 8 is Decent Work and Economic Growth Impact: Economic growth that provides

Decent jobs can influence population mobility and urbanization, leading to changes in demographic trends and family structures. Goal 11 is Sustainable Cities and Communities Impact: Urbanization can present challenges such as overcrowding and inadequate services. Sustainable urban planning can lead to manage population density and enhance living conditions. Goal 10 is Reduced Inequality Impact: Addressing inequalities within and between Countries can help manage population pressures by ensuring that all groups have access to resources and opportunities. Goal 13 is Climate Action Impact: Climate change impacts migration patterns and can exacerbate population pressures. Sustainable practices can help communities to resilience these effects of climate change. Sustainable practices can strengthen communities' ability to cope with and recover from climate change impacts. Goal 17 is Partnerships for the Goals Impact: Global cooperation is essential for addressing population challenges. Collaborative efforts can enhance resource allocation and share best practices across regions. Conclusion the SDGs provide a comprehensive approach to addressing population challenges by focusing on health, education, gender equality, and economic growth. By tackling these interconnected issues, the SDGs aim to create a sustainable future for all. Relation between the Sustainable Development Goals (SDGs) and population The Sustainable Development Goals (SDGs) are positively related with the population dynamics, as they address various factors that influence population growth, distribution, and well-being given below.

1. Demographic Transition the SDGs promote conditions that can lead to demographic transition, where societies transform from high birth and death rates to lower rates. This transition made possible due to industrial, medical and agricultural revolution. encourage this transition.
2. Health and Reproductive Rights (SDG 3) Access to healthcare, including reproductive health services, is vital for managing population growth. Empowering individuals to make informed choices about Family size direcctly impacts population rates.
3. Education and Empowerment (SDG 4 and SDG 5) Education, particularly for women and girls, is linked to lower fertility rates and improved health outcomes. Empowerment enables women to make informed reproductive choices, influencing the population dynamics.
4. Economic Growth and Employment (SDG 8) Economic Development can affect migration patterns and urbanization, impacting population distribution. Decent work opportunities can help manage population pressures in urban areas.
5. Urbanization and Sustainable Communities (SDG) 11) Rapid urbanization due to population growth can strain resources. Sustainable urban planning is essential for managing population density and ensuring adequate infrastructure and services.
6. Inequality and Social Cohesion (SDG 10) Reducing inequality ensures that all populations, including marginalized groups, have access to education, healthcare, and economic opportunities, which can stabilize population growth.
7. Climate Change and Migration (SDG 13) Climate change can lead to forced migration, impacting population distribution. Sustainable practices help communities adapt and mitigate the effects of climate change.
8. Partnerships for Sustainable Development (SDG 17) Global cooperation is essential for addressing population challenges. Collaborative efforts can enhance resource allocation and share best practices across regions. The SDGs collectively address the multifaceted issues related to population, recognizing that sustainable development requires Planning of population dynamics using overall balance development. sustainability.

By achieving these goals, countries can better manage them. population challenges and promote a more equitable and sustainable future.

References

Agarwala, S.N. (1962). Age at marriage in India. Kitab -Mahal Publisher, Allahabad.

Asha Bhende Tara Kanitkar (2008): Principals of Population studies. Himalaya Publishing house, Mumbai.

Bhagat. (2002). *Early Marriages in India: A Socio-geographical Study.* New DElhi: Rajat Publications.

BLACKER, C.P. (1949): "Stages of population growth", The Eugenics Review, 39 (3), 88-101.

CHESNAIS, J.C. (1992): "The demographic transition", Oxford University Press.

Coal, Ansley J. (1971) "Age Pattern of marriage," Population Studies. 25, 193-214.

COALE, A.J. and HOOVER, E.M. (1958): "Population growth and economic development in low-income countries", Princeton, N.J., 10-13.

DAVIS, K. (1945): "The world demographic transition", Annals of the American Academy of Political and Social Science, 273 (Jan.), 1-11.

Demographic Health Survey (DHS)

Goyals, R.P. Nuptiality trends and Differentials. Population of India, Country monograph series no. 10, ESCAP, United nations Pp. 109-133.

Hajnal J. (1953). Age at marriage and population studies, 7(2) 111-136.

Kapadia K. M. 1966. Marriage and family in India, Oxford University Press. Pp. 395.

LANDRY, A. (1934): "La révolution démographique : études et essais sur les problèmes de population", (Paris, Librairie Sirey).

Malaker, C. R. (1978), "Study of Indian Nuptiality", Ph.D. dissertation, Calcutta: Indian Statistical Institute.

International Institute of Population Sciences. Mumbai. IPS

NOTESTEIN, F.W. (1945): "Population: the long view", in E. SCHULTZ (ed.) Food for the world (Chicago, III: University of Chicago Press), 36-57.

Pandey, Arvind (1984), "Determinants of Age at Marriage for Females in India", unpublished paper, Mumbai, India: International Institute for Population Sciences.

Pathak, K. B. (1980), "Law and Age at Marriage for Females in India", in Journal of Social Work, vol. 40, no. 4, pp. 407-415, New Delhi: SAGE.

Sklar J.L. 1974. The role of Marriage behavior in Demographic Transition: The case of eastern Europe around 1900'. Population studies. xxviii2. Pp. 231-47.

The Economics of Population and food supplies: economic problems of population change", Proceedings of the 8th International Conference of Agricultural economists, London, 1953, 13-31.

United Nation

Vijaya Khairkar (2007). Sociao- economic of Migration. Diamond Publication, Pune. pp 150.

Vijaya Khairkar (2013). " Loksankhya Shastra" by A.G. Publication, Pune . PP 150.

World Fertility Surveys (WFS)

MAP 1

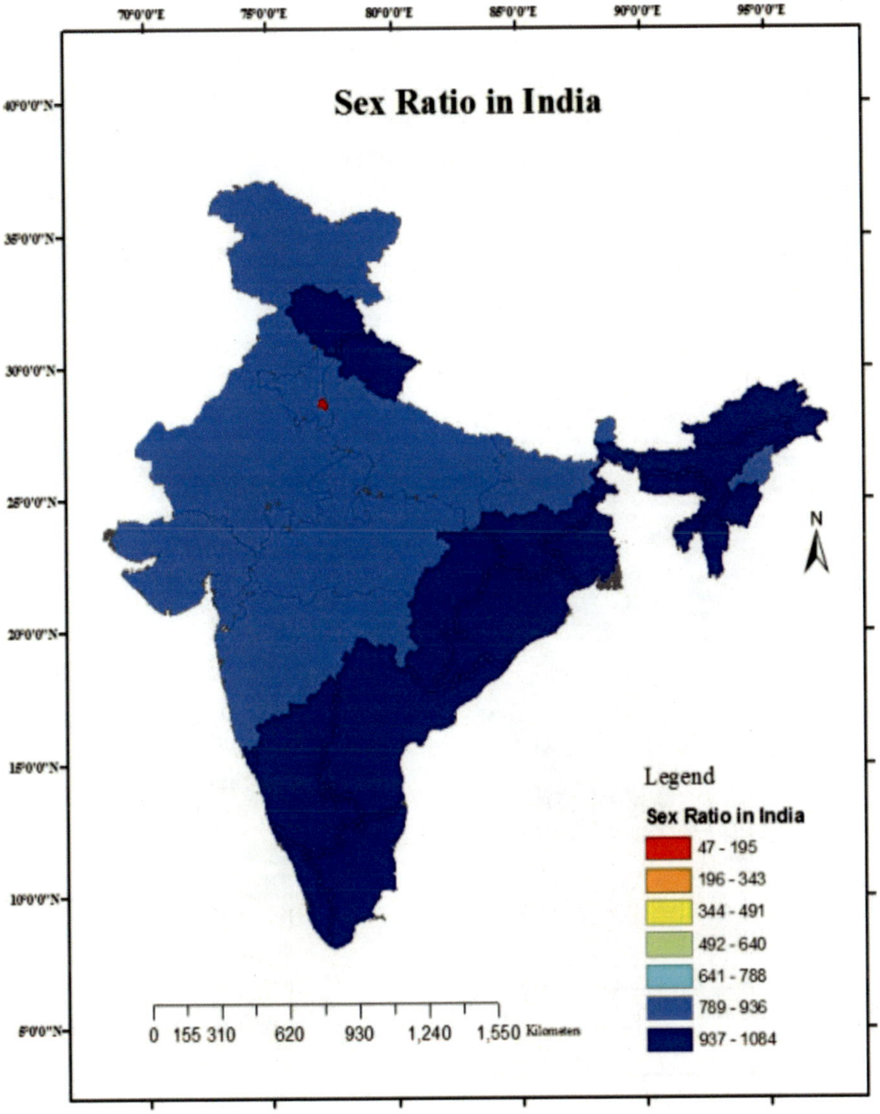
Sex Ratio in India
Legend
Sex Ratio in India
47 - 195
196 - 343
344 - 491
492 - 640
641 - 788
789 - 936
937 - 1084
0 155 310 620 930 1,240 1,550 Kilometers
N

MAP 2

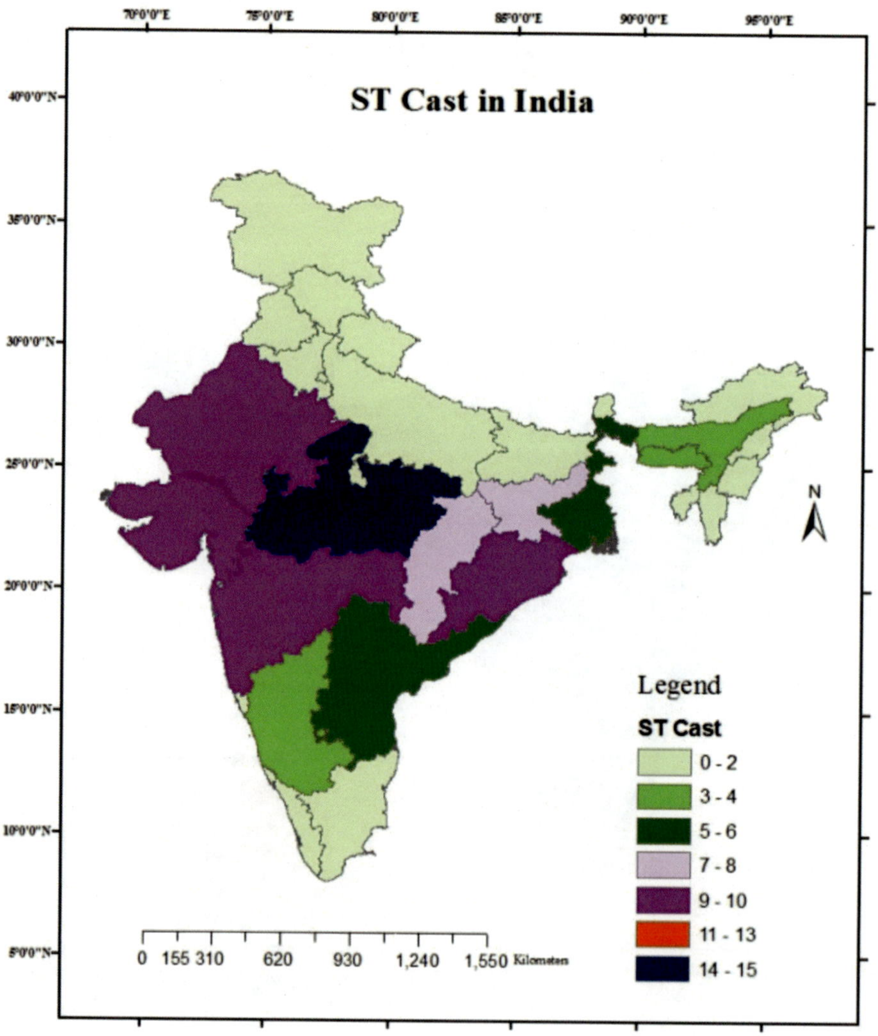
ST Cast in India
70°0'0"E
75°0'0"E
80°0'0"E
85°0'0"E
90°0'0"E
95°0'0"E
40°0'0"N
35°0'0"N
30°0'0"N
25°0'0"N
20°0'0"N
15°0'0"N
10°0'0"N
5°0'0"N
N
Legend
ST Cast
0 - 2
3 - 4
5 - 6
7 - 8
9 - 10
11 - 13
14 - 15
0 155 310 620 930 1,240 1,550 Kilometers

MAP 3

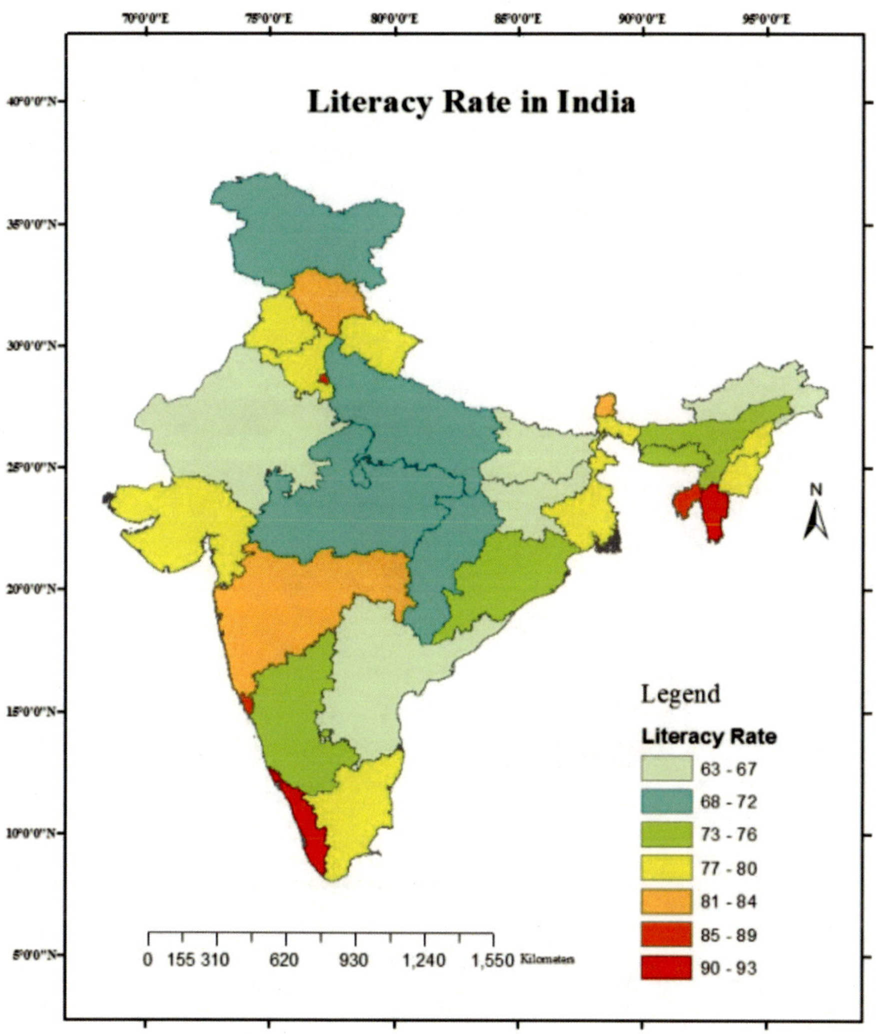
70°0'0"E
75°0'0"E
80°0'0"E
85°0'0"E
90°0'0"E
95°0'0"E
40°0'0"N
35°0'0"N
30°0'0"N
25°0'0"N
20°0'0"N
15°0'0"N
10°0'0"N
5°0'0"N
Literacy Rate in India
N
0 155 310 620 930 1,240 1,550 Kilometers
Legend
Literacy Rate
63 - 67
68 - 72
73 - 76
77 - 80
81 - 84
85 - 89
90 - 93

MAP 4

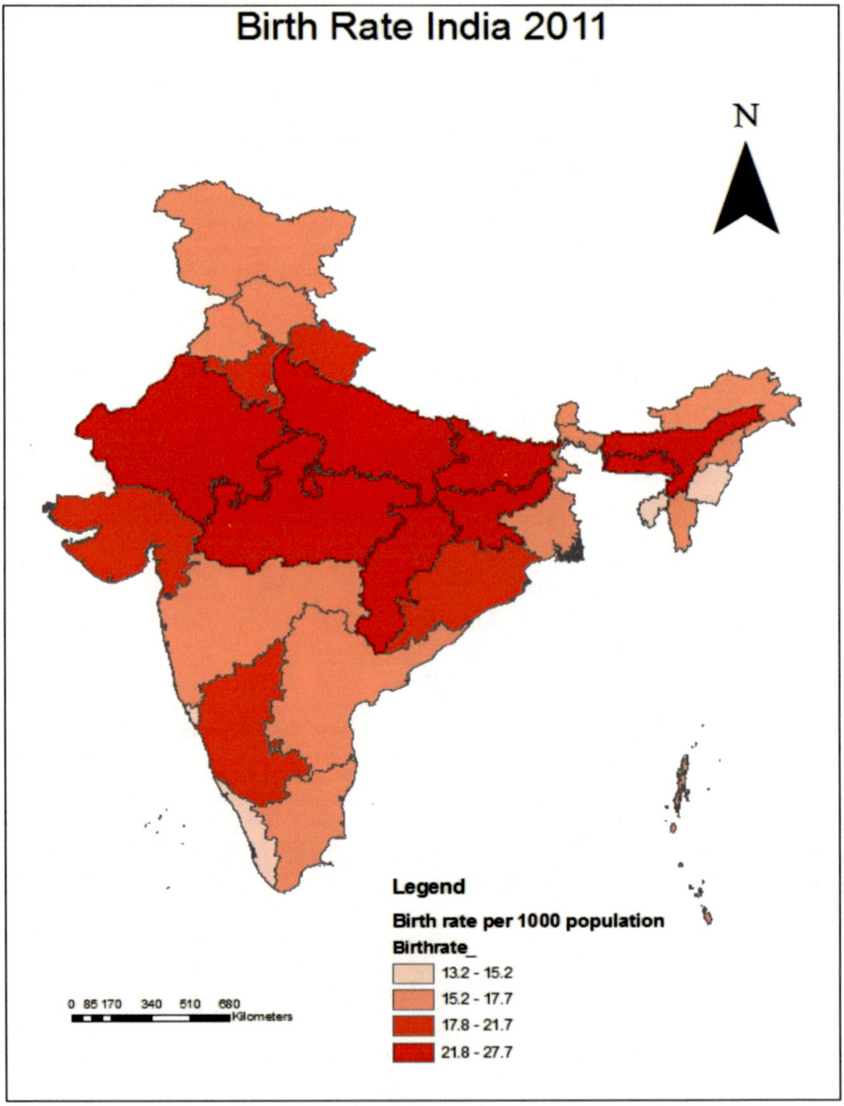
Birth Rate India 2011
N
Legend
Birth rate per 1000 population
Birthrate_
13.2 - 15.2
15.2 - 17.7
17.8 - 21.7
21.8 - 27.7
0 85 170 340 510 680
Kilometers

MAP 5

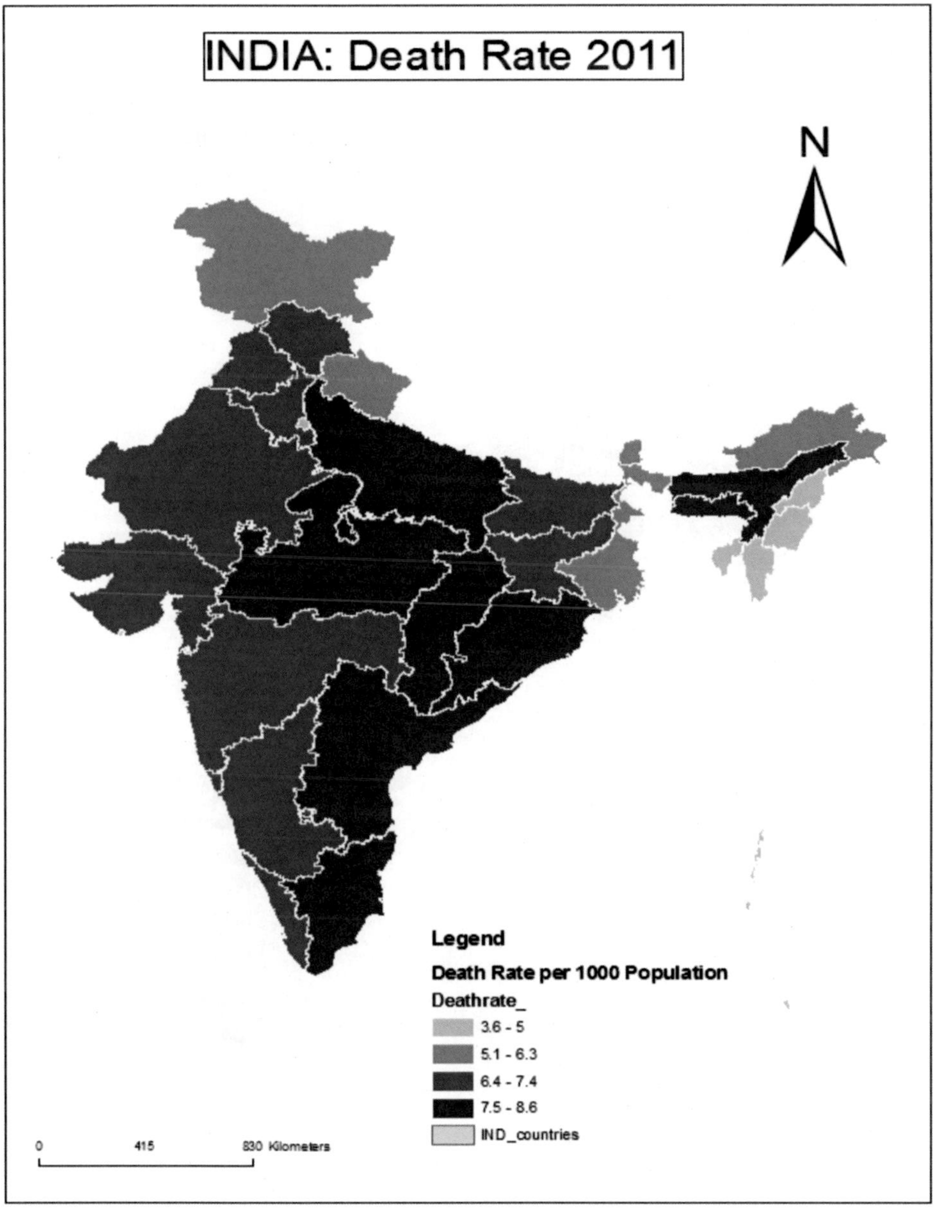
INDIA: Death Rate 2011
N
Legend
Death Rate per 1000 Population
Deathrate_
3.6 - 5
5.1 - 6.3
6.4 - 7.4
7.5 - 8.6
IND_countries
0
415
830 Kilometers

MAP 6

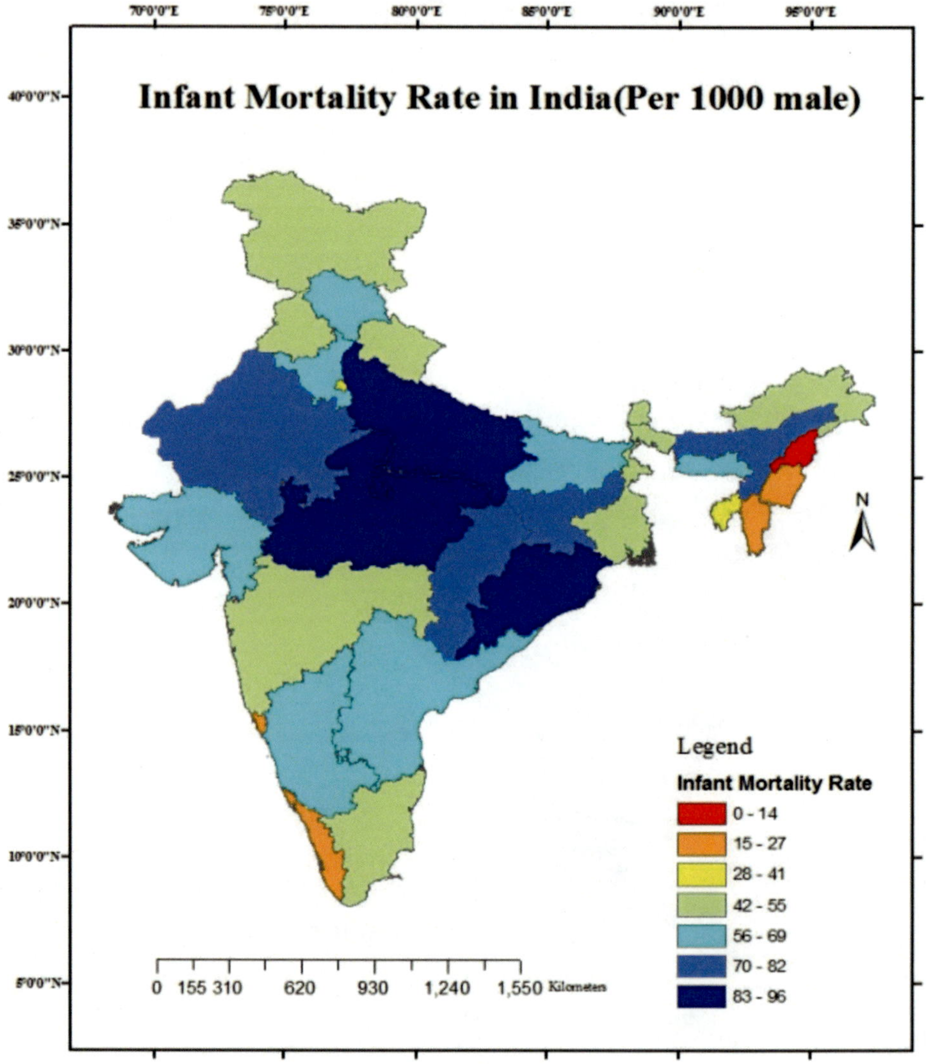
Infant Mortality Rate in India(Per 1000 male)
70°0'0"E
75°0'0"E
80°0'0"E
85°0'0"E
90°0'0"E
95°0'0"E
40°0'0"N
35°0'0"N
30°0'0"N
25°0'0"N
20°0'0"N
15°0'0"N
10°0'0"N
5°0'0"N
N
Legend
Infant Mortality Rate
0 - 14
15 - 27
28 - 41
42 - 55
56 - 69
70 - 82
83 - 96
0 155 310 620 930 1,240 1,550 Kilometers

MAP 7

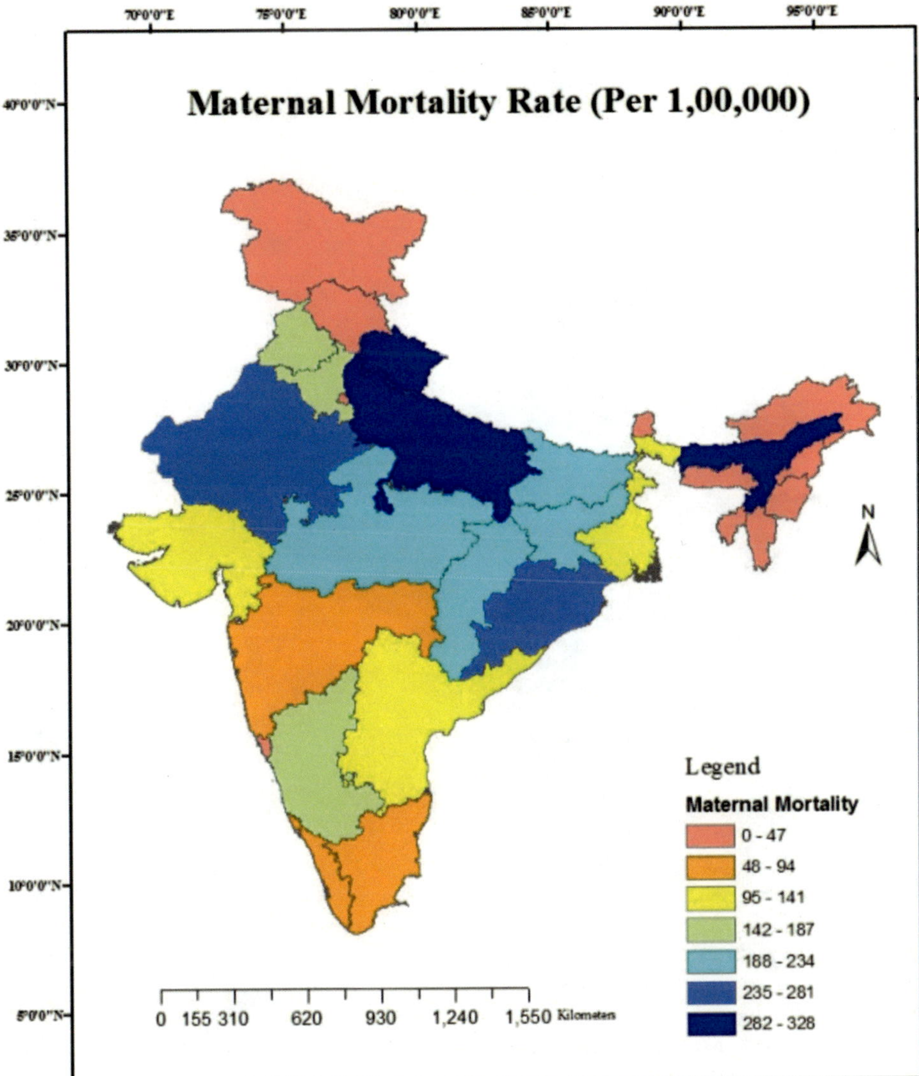
70°0'0"E
75°0'0"E
80°0'0"E
85°0'0"E
90°0'0"E
95°0'0"E
40°0'0"N
35°0'0"N
30°0'0"N
25°0'0"N
20°0'0"N
15°0'0"N
10°0'0"N
5°0'0"N
Maternal Mortality Rate (Per 1,00,000)
N
Legend
Maternal Mortality
0 - 47
48 - 94
95 - 141
142 - 187
188 - 234
235 - 281
282 - 328
0 155 310 620 930 1,240 1,550 Kilometers

MAP 8

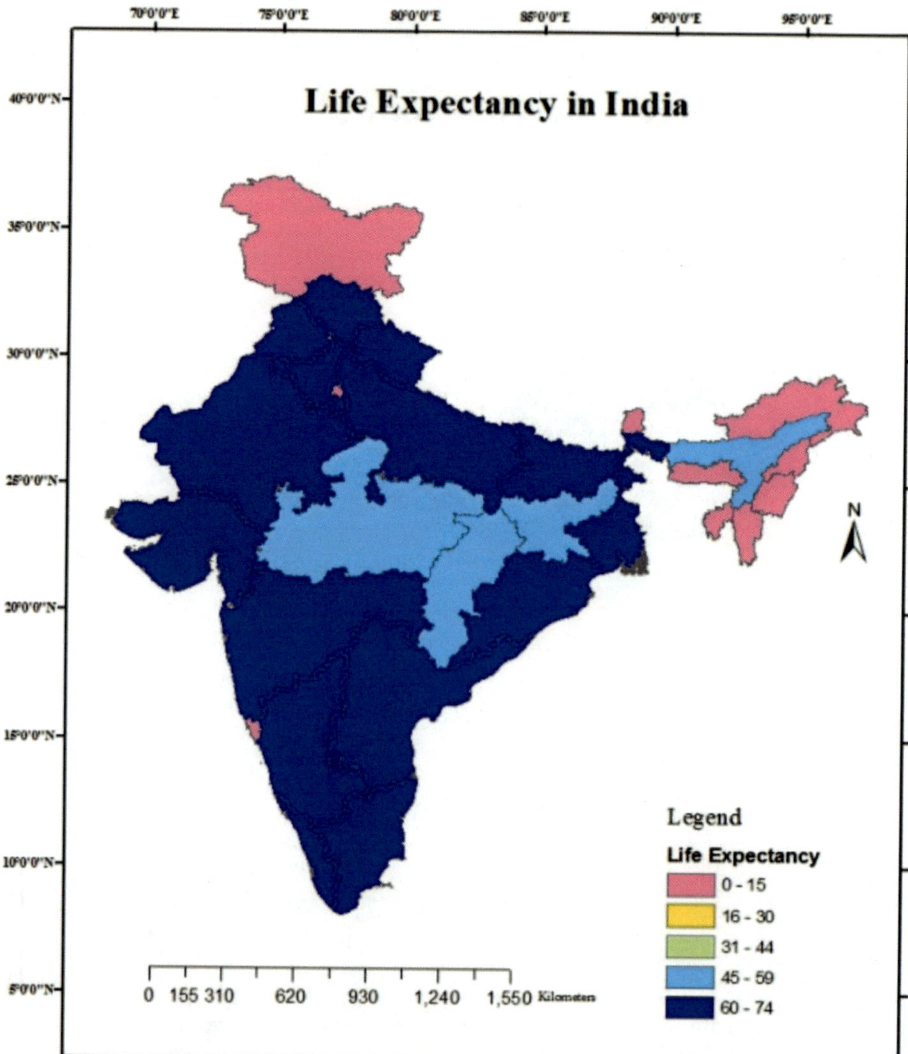
Life Expectancy in India
Legend
Life Expectancy
0 - 15
16 - 30
31 - 44
45 - 59
60 - 74
N
0 155 310 620 930 1,240 1,550 Kilometers

MAP 9

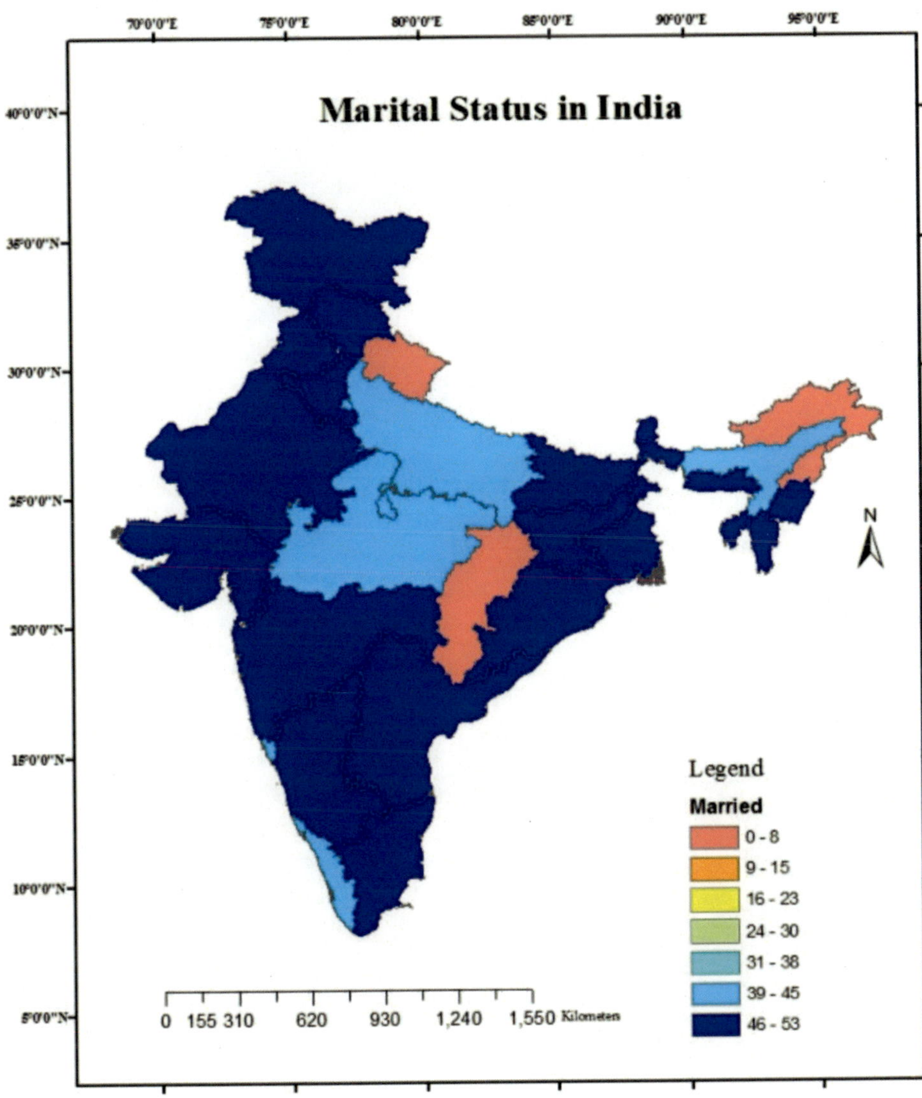
Marital Status in India
70°0'0"E
75°0'0"E
80°0'0"E
85°0'0"E
90°0'0"E
95°0'0"E
40°0'0"N
35°0'0"N
30°0'0"N
25°0'0"N
20°0'0"N
15°0'0"N
10°0'0"N
5°0'0"N
N
Legend
Married
0 - 8
9 - 15
16 - 23
24 - 30
31 - 38
39 - 45
46 - 53
0 155 310 620 930 1,240 1,550 Kilometers